Nature Library

ASTRONOMY

Nature Library

ASTRONOMY

Storm Dunlop

Viscount Books

Artists

Max Ausell, A. C. Bishop, Nigel Chamberlain, Jim Channell, Ralph S. Coventry Associates, Gordon Davies, Jon Davies, Phil Davies, Design Practitioners, Keith Duran, Tony Gibbons, Harry and Gwen Green, Graham Gurr, Terry Hadley, W. R. Hamilton, David A. Hardy, Hayward Art Group, Bill Hobson, David Lewis Artists, Oxford Illustrators, Jack Pelling, John Smith, Mick Tregenza, Tudor Art Agency Ltd, Brian Watson, Whitecroft Designs, A. R. Woolley.

The Preliminary Geological Terrain Map of Mercury on pp. 20-1 is published by permission of N. J. Trask and J. E. Guest, from *Journal of Geophysical Research*, vol. 80, no. 17, pp. 2461-77, 1975. The drawings on page 63 are published by permission of Professor J. Toomre, University of Colorado.

This edition published by Viscount Books 1984.
Prepared by Newnes Books,
a division of The Hamlyn Publishing Group Limited,
84-88 The Centre, Feltham, Middlesex, England
and distributed for them by
The Hamlyn Publishing Group Limited,
Rushden, Northants, England.

ISBN 0 600 35661 2

Printed in Italy.

Photographic acknowledgements

ARIZONA OFFICE OF TOURISM 39 top right; ASSOCIATION OF UNIVERSITIES FOR RESEARCH IN ASTRONOMY, INC., KITT PEAK NATIONAL OBSERVATORY, TUCSON 55; JULIAN BAUM frontispiece; BIBLIOTECA MARUCELLIANA, FLORENCE 12 top; BRITISH MUSEUM, LONDON 74 right; CALIFORNIA INSTITUTE OF TECHNOLOGY AND THE CARNEGIE INSTITUTION OF WASHINGTON, reproduced by permission from THE HALE OBSERVATORIES, PASADENA 50, 52 top, 52 bottom, 53 top left, 56-7, 77; WILLIAM COBLEY 24 bottom; STORM DUNLOP 71 bottom; MARY EVANS PICTURE LIBRARY, LONDON 11 bottom, 13 bottom; FOCAL POINT, PORTSMOUTH 51, 53 top right; PETER GILL 8, 73; JOHN GREEN 9; HALE OBSERVATORIES, PASADENA 14 centre, 52 top; HAMLYN GROUP 10, 11 top, 13 top left, 23 top, 37, 39 left; HANSEN PLANETARIUM, SALT LAKE CITY 74 left; MICHAEL HOLFORD LIBRARY, LOUGHTON 46 top; STEPHEN LARSON/ UNIVERSITY OF ARIZONA 43; ROBERT MCNAUGHT 60; DAVID MALIN/ROYAL OBSERVATORY, EDINBURGH 62 top left; NASA, WASHINGTON D.C. 18, 19 top left, 19 bottom, 20, 22 top, 22 bottom, 23 centre, 23 bottom, 24 centre, 29 centre, 29 bottom, 30 top left, 30 top right, 30-1, 32, 34 top, 34 bottom (and inset), 35, 36, 38, 40 top, 40 bottom, 41, 42 top, 42 bottom, 45, 47, 53 centre, 62 top right, 64 right, 70, 71 top; NATIONAL PORTRAIT GALLERY, LONDON 13 top right, 13 centre; PHOTRI, ALEXANDRIA, VIRGINIA 33, 46 bottom, 67 bottom; ROYAL OBSERVATORY, EDINBURGH 6-7, 14 top, 54, 66; SCIENCE PHOTO LIBRARY, LONDON: Dr J. Lorre 59, NASA 69 bottom left, X-ray Astronomy Group, University of Leicester 61, 69 top; SCIENCE AND ENGINEERING RESEARCH COUNCIL, SWINDON 69 centre right; SMITHSONIAN INSTITUTION, WASHINGTON D.C. 67 centre; SPACE FRONTIERS, HAVANT 48-9; R. G. STROM, WESTERBORK RADIO OBSERVATORY, HOLLAND 69 bottom right; HANS VEHRENBERG 27; VON DEL CHAMBERLAIN 49; DR JAMES D. WRAY 62 bottom, 64 left.

Contents

Introduction

Astronomy has always been one of the most fascinating sciences. It may have begun in superstition, with attempts to fortell what the future held for individuals or nations, but throughout its long history – and it is the oldest of the natural sciences – it has gradually explained many of the marvels of the heavens. In this respect astronomers are, perhaps, very lucky people, as they have the whole Universe to study. However, anyone can appreciate the beautiful (and sometimes strange) pictures obtained with large telescopes and spacecraft, as well as understand something about the wonderful objects which have been discovered.

There is a mistaken idea that astronomy is complicated, and that it involves a lot of mathematics and expensive equipment. For professional astronomers it may be so, but there are many thousands of ordinary people all around the world who find it an absorbing hobby. Some have no equipment at all, except their eyes, and many more are quite useless at mathematics, but this does not stop them from gaining great enjoyment, and if they should wish, of making observations of considerable scientific importance.

It is hoped that this book will serve as an introduction to this enthralling science, and perhaps encourage some of its readers to take their studies further by observing the night sky.

The Sun and stars

Everyone recognises the apparent motion of the Sun in the sky, rising in the east and setting in the west, but it is surprising how few people realise that the stars also do the same, due to the Earth's rotation. Ancient peoples, however, being untroubled by city lights, were quite aware of the motion of the stars, and of how they appeared to circle around the pole of the heavens. Once agriculture came to be practised, and a knowledge of the reckoning of the seasons became all-important, true study of the motions of the Sun and Moon was started.

In the course of a year, as the Earth passes round the Sun, the stars that can be seen alter with the seasons and the Sun itself appears to move against the background of the stars. The path it traces is called the **ecliptic.** Being tilted to the Earth's equator, the ecliptic carries the Sun high in summer and low in winter. Due to the inclinations of their orbits the Moon and planets appear to swing either side of the ecliptic, within the band of the 12 constellations of the Zodiac.

Or at least they did, once upon a time. Unfortunately for astrologers (and their superstitious believers), the position of the Earth's rotational axis is continuously changing relative to the stars, mainly due to the gravitational effects of the Sun and Moon. As a consequence the position of the ecliptic has changed over the years, and the 'star-signs' are no longer valid. The Sun now spends more time in the constellation of Ophiuchus than it does in neighbouring Scorpius, and the Moon and planets can wander into all sorts of constellations.

The natural divisions of time are those of the day and the year, both of which are defined by the position of the Sun in the sky. All calendars have been attempts to reconcile these two lengths of time, sometimes with

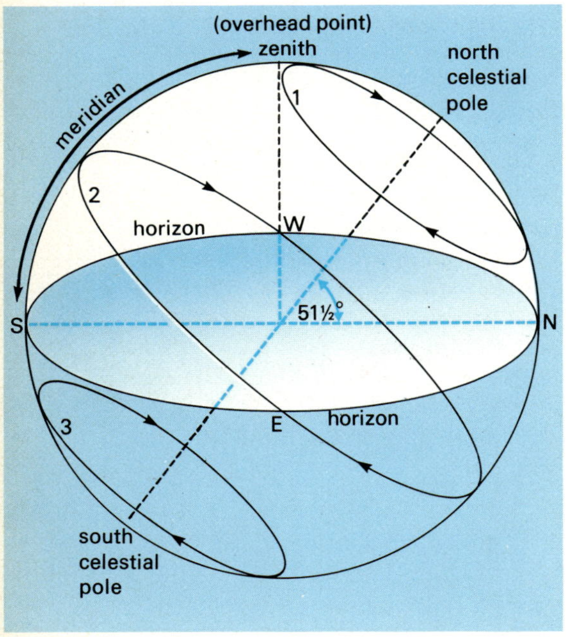

Left: **the observer's latitude determines which stars can be seen and those that never set. From the Equator (in theory) every star is visible.**

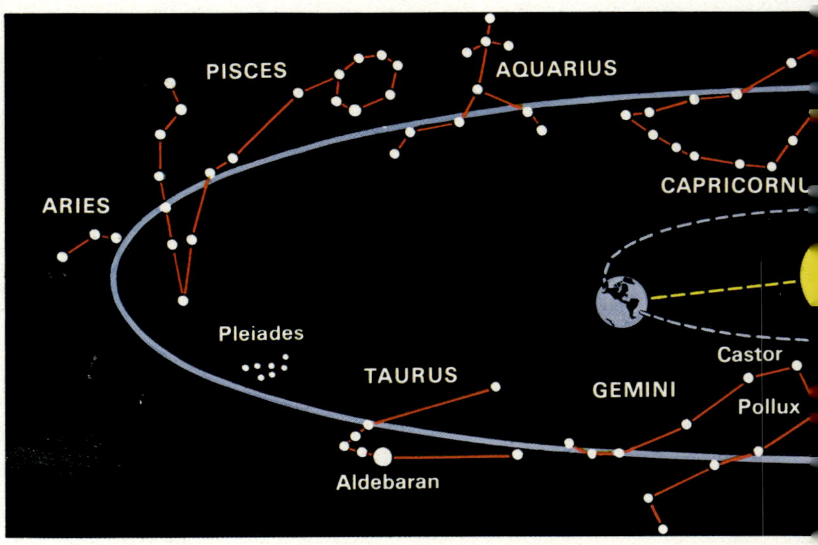

Below: **the apparent motion of the stars during a two-hour exposure with the camera pointing towards the North Pole and Polaris (the bright trail just left of centre).**

that of the lunar motion around the Earth as well. But each has been a compromise, because the year consists of 365·242 days (approximately) not a nice round 365 or 365¼. Astronomers have five different lengths of 'month', depending upon how it is measured, and there are just as many 'years'. The length of a day is also gradually changing because the Earth is being slowed down by the action of the Moon. As time is now measured by the unchanging rate of atomic clocks, this alteration can be measured. Many people are puzzled by the apparently silly fact that occasionally a second is 'lost' from clock time around the Earth. It is, however, only in this way that we can keep artificial time in close step with the ever-changing time marked out by the Sun.

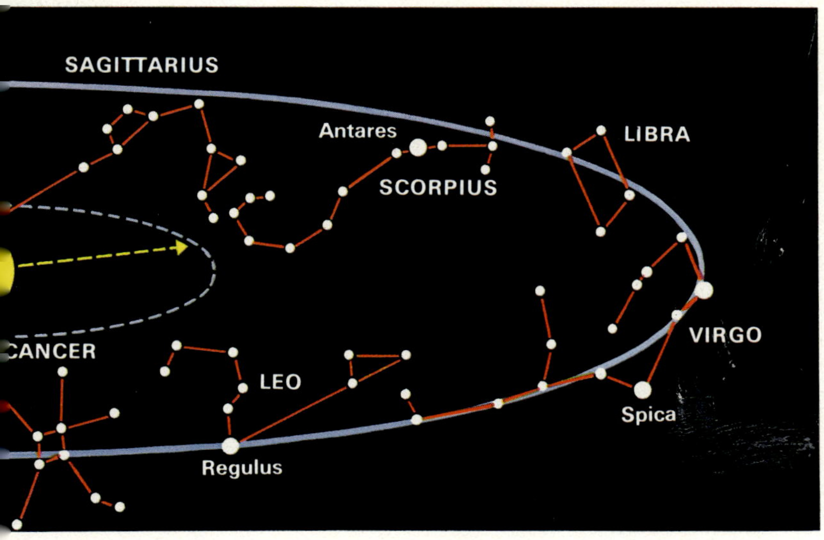

Above: Stonehenge, like many other megalithic sites, was probably used to follow the movements of the Sun and Moon, thus marking out the seasons of the year.

Left: as the Earth moves round its orbit, the Sun appears to travel along the ecliptic, and through the various constellations of the Zodiac.

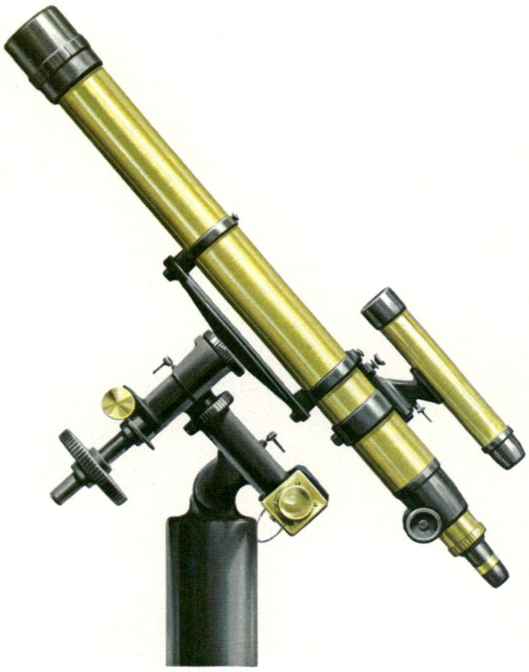

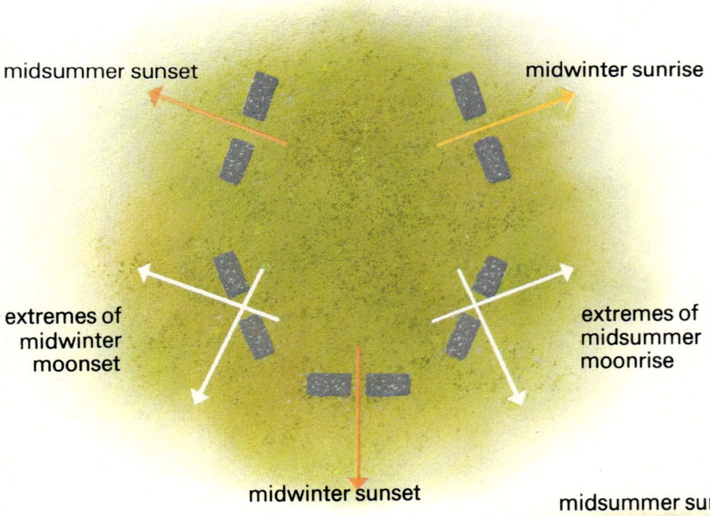

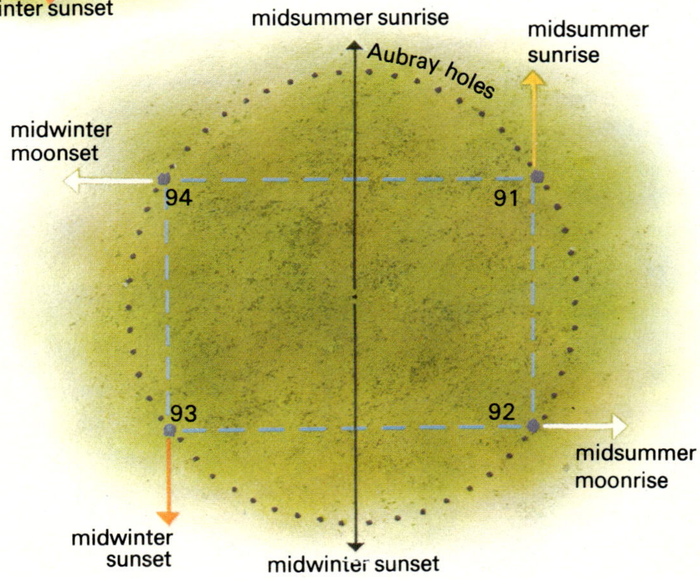

This telescope is pointing towards the North Pole. Its mounting has one axis parallel to that of the Earth, allowing easy compensation of the daily rotation.

The main alignment at Stonehenge is towards midsummer sunrise, but there are many others. The Aubrey holes may have been used as a form of eclipse predictor.

The motion of the planets

The motions of the planets against the stars are far more complicated than those of the Sun and Moon, and it was many centuries before they were understood. The trouble arises, of course, from the fact that the Earth itself does not appear to move at all, while the planets seem to go through a complicated dance around it. Mercury and Venus are always close to the Sun in the sky, while the other planets can not only be on the other side of the heavens, but also go through loops and move backwards against the stars.

Although the idea of a **heliocentric** system (centred on the Sun) was suggested by certain early Greek astronomers, notably by Aristarchos in about 260 BC, the idea generally accepted was the **geocentric** one (i.e. in which Sun, Moon and planets revolved around the Earth) particularly promoted by Aristotle. Such a system could not in itself account for the apparent motions, and it was Hipparchos who suggested that the planets described small circles, the centres of which themselves revolved around the Earth. This general idea was copied and refined in about AD 100 by the Alexandrian astronomer Ptolemy, after whom the general scheme is usually known.

The Ptolemaic system reigned supreme for many hundreds of years, until with greater accuracy of positional measurements it became obvious that it could not account for the exact motions of the planets, and more and more 'refinements' had to be added, making the scheme ever more complicated. It had always been appreciated that a heliocentric system could account for the observed movements, but it was not until Nicolas Copernicus finally published his theory in 1543 that it was seriously considered. The theory was immediately used to prepare new tables of planetary positions. Despite its success, and partly for reasons of religious dogma, the 'new' theory was not accepted immediately. In particular the greatest of pre-telescopic observers, the Dane Tycho Brahe, proposed in 1583 that all the planets except the Earth had orbits centred on the Sun, but that the Sun itself (and consequently all the other planetary bodies) still went round the Earth.

What we would now accept as observational proof of the correctness of the Copernican theory was obtained by Galileo in 1609-10, and at almost exactly the same time Johann Kepler found the true solution. Up to that time everyone, including Copernicus, had made the assumption, partly based on ideas of 'perfection', that the planets moved only in circles. Kepler came to realise that this was not the case, but that the motions could be explained exactly if the planets moved in elliptical orbits in an otherwise Copernican system. In 1619 he published the last of his three laws, finally relating the distance of a planet to its orbital period around the Sun.

Below: **the early Greek theory that all the heavenly bodies, including the Earth and the Sun, circled an invisible 'Central Fire'.**

Opposite top left: **in the Ptolomaic view, a planet was carried in a small circle, the *epicycle* around a larger circle, the *defferent*, centred on the Earth.**

Opposite centre: **in the heliocentric universe of Copernicus, all the planets, including the Earth, orbited around the Sun, while the stars were 'fixed', and at a great distance.**

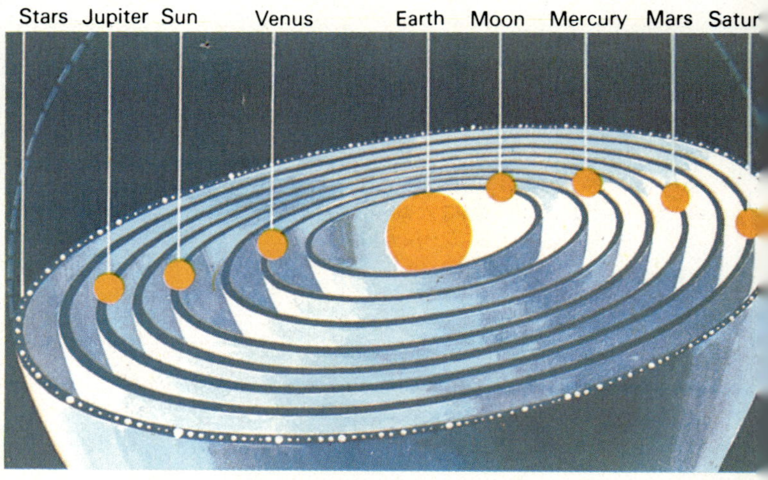

Above: **in the Aristotelian universe concentric, transparent, crystal spheres carried all the heavenly bodies around the Earth, which was motionless at the centre.**

Above: **the type of looping path followed at times by the outer planets. The motions of Mercury and Venus are even more complicated.**

Right: **the Ptolomaic universe, which was centred on the Earth, around which all the other heavenly bodies were thought to rotate.**

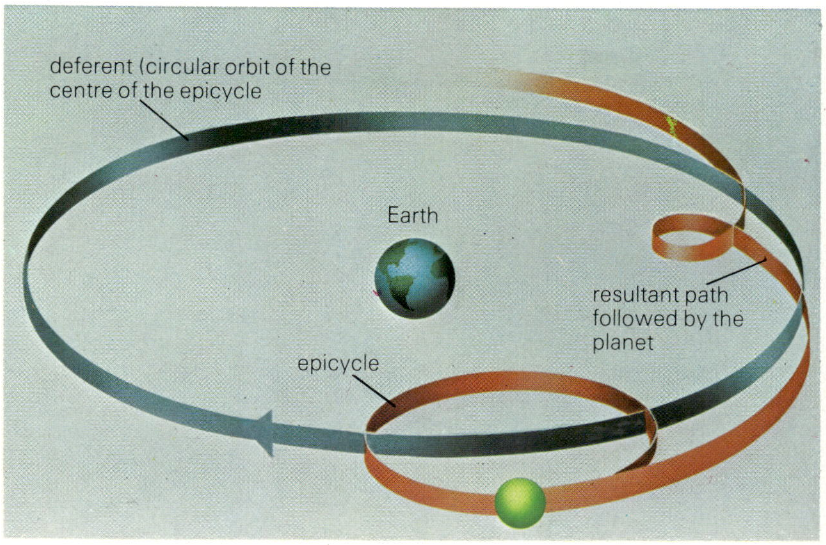

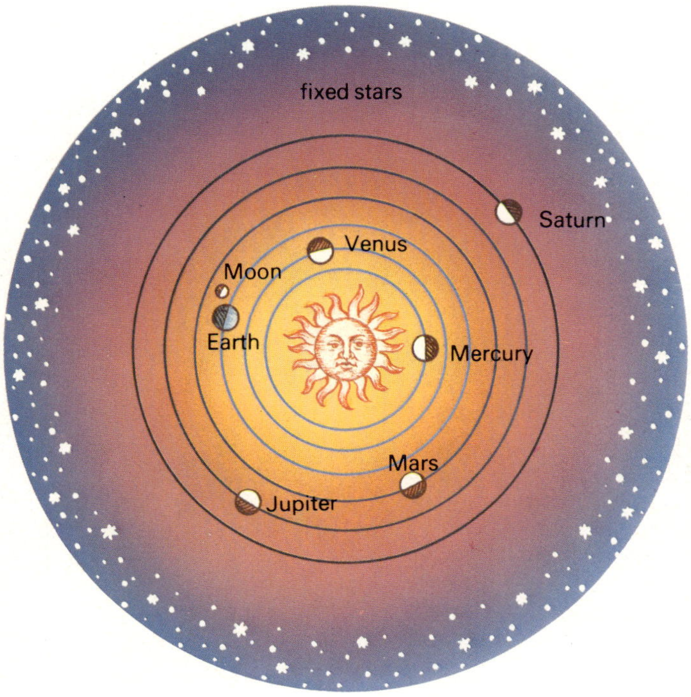

Top right: **the Polish astronomer, Nicholas Copernicus, finally showed that the Earth did not lie at the centre of the Universe.**

Right: **it was the great difficulty in accounting for the motions of the planet Mars that led Kepler, after many years of effort, to his theory of planetary motion.**

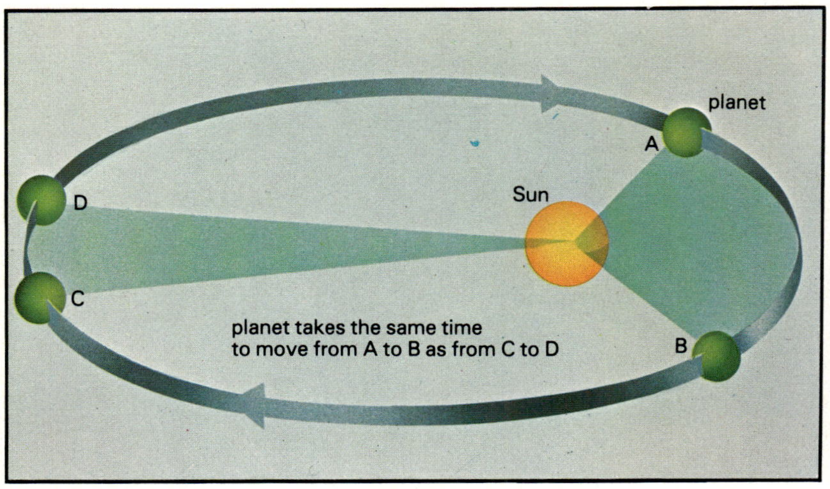

Far left: **the speed of a planet's movement in its orbit, as Kepler showed, is not constant, but is fastest when close to the Sun, and slower when further away.**

Left: **Johannes Kepler, whose three laws of planetary motion enabled the future positions of the planets to be accurately predicted.**

The birth of modern astronomy

By the late 16th century increasing attention was being paid to several important problems. Apart from the understanding of planetary motions and the wish to predict the orbits, there was a desire for more detailed knowledge of the positions of the stars. The general movement led in 1580 to the establishment of the first true astronomical observatory, built on the island of Hveen for Tycho Brahe, under the patronage of the Danish king, Frederick II. The telescope had not yet come into use, but Tycho's large instruments were the most accurate that could be devised, and the precise measurements that he obtained were to prove crucial in the development of Kepler's theories of planetary motion.

A little later Galileo Galilei seized upon the idea of the newly-invented telescope and made a series of discoveries which completely changed the nature of astronomy. The mountainous nature of the Moon, the phases of Venus, sunspots (although in this Scheiner was probably first), the rotation of the Sun, and the fact that the Milky Way actually consisted of countless stars, were just a few of his contributions. More important still, he discovered the four major satellites of Jupiter (still called the Galilean satellites), which proved unquestionably that bodies could orbit something other than the Earth. These and other discoveries demonstrated that the heavens were not changeless.

In later years many astronomers followed up these earlier discoveries. Giovanni Cassini, for example, finally established the true scale of the Solar system when he managed to measure the distance of Mars, and by extension, find that of the Sun. His measurements of the positions of the Galilean satellites were later used by Olaus Roemer in determining the speed of light.

Kepler revealed the elliptical nature of the paths which the planets followed, but the explanations of *why* these should be ellipses had to await Sir Isaac Newton's development of the theory of universal gravitation. Among other things this introduced the most important concept of 'mass' – the quantity of matter within a body – which, unlike weight, does not vary. It was the application of Newton's theory that enabled Edmond Halley to predict the return of the comet

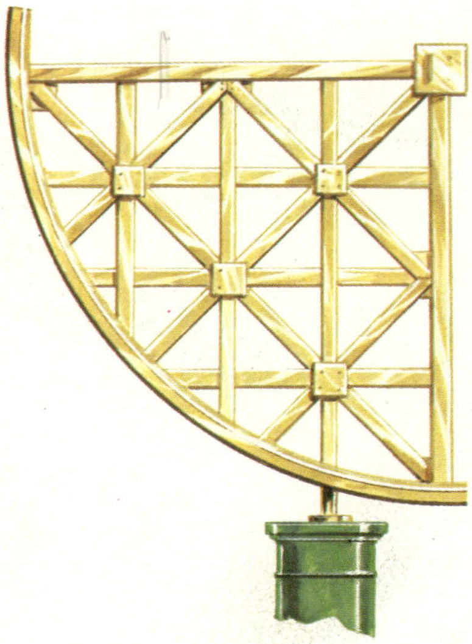

Left: **Galileo Galilei, the first astronomer to use a telescope, made many outstanding contributions to the understanding of the heavens.**

Right: **Tycho Brahe's sextant** *(above)* **did not have telescopic sights, and even Galileo's most powerful telescopes did not magnify more than about 30 times.**

Giovanni Cassini, Italian-born, but later Director of the Paris Observatory, made many important observations of the planets and discovered several satellites of Saturn.

Tycho Brahe's observatory of Uranieborg was designed to house many different instruments, most of them very large and of the highest quality for the time.

One of Newton's tiny reflecting telescopes, the first of which was built in 1668, after the principle was suggested by the Scot, James Gregory.

Above: **the theory of gravitation developed by Sir Isaac Newton at last solved the age-long question of the motion of the planets.**

which now bears his name, thus solving another outstanding problem.

It was Newton, too, who made the discovery that white light could be spread into a spectrum. He made an error here, believing his results to show that a refracting telescope could not be made free from the troublesome defect of chromatic aberration. But the error caused him to build the first reflecting telescope – the precursor of the giants of today.

Reflecting telescopes became all-important with their development by Sir William Herchel, who became famous for his discovery in 1781 of the planet Uranus, the first new planet that had not been known in antiquity. Herschel went on to lay the foundations for the proper study of the stars and of the Galaxy as a whole.

Left: **Sir William Herschel– perhaps the greatest observational astronomer who has ever lived– contributed greatly to the development of the telescope.**

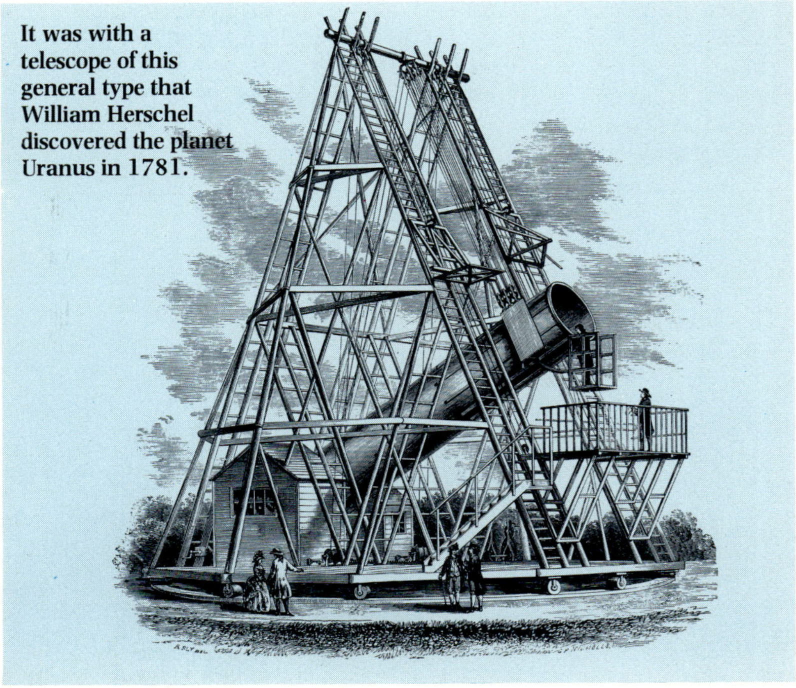

It was with a telescope of this general type that William Herschel discovered the planet Uranus in 1781.

Herschel's greatest telescope, the 'forty foot' (12·2m), with a mirror 48 inches (1·22m) in diameter, was the largest in the world when built in 1789.

Distance and time

Any astronomer has to become used to dealing with vast distances and long spans of time, and these frequently prove very difficult to understand. Astronomers find it just as awkward to visualize large numbers as anyone else, but they have adopted the trick of using various different units of measurement to make the dimensions easier to comprehend. For most planetary objects we can conveniently use kilometres, even though the diameter of the Earth is 12 756km, and that number alone is difficult enough to visualize. The average distance of the Moon is rather more than 384 400km, but, although we may just about accept it, that figure is rather meaningless. Perhaps it would be better to say that it takes slightly more than $1\frac{1}{4}$ seconds for light, or radio waves, to cross that distance. (Anyone who has listened to a recording of the Apollo Moon landings will probably remember the pauses between questions and answer, and may have experienced for themselves the slight delays that occur on intercontinental telephone calls requiring the use of more than one relay satellite in geostationary orbit 35 900km above the Earth.)

For distances within the Solar System, the **astronomical unit** (the average distance of the Earth from the Sun, 149 597 870km) is usually employed. So the average distance of Pluto, the most distant planet, is 39·4 astronomical units. But we could also say that it takes light from the Sun just over 8 minutes (actually 8 mins 19 secs) to cover one astronomical unit, and about $5\frac{1}{2}$ hours to reach Pluto.

When we discuss the distances of the stars we must again use a different unit, because even the nearest star, Proxima Centauri, is at a distance of about 270 000 astronomical units. We may conveniently express this as about 4·26 **light-years.** (Note that a light-year – the distance covered in one *year* by light travelling at 299 792·458km per *second* – is a measurement of distance, not of time.)

Left primitive _Australopithecus_ was Earth's highest lifeform when this light started its journey from the Andromeda Galaxy _(above)_ 2·2 million light years away.

Even the nearest clusters of galaxies are so distant that their light takes more than 150 million years to reach us. At that time _Archaeopteryx_ _(below left)_ the earliest bird-like creature existed on Earth, as well as dinosaurs like _Stegosaurus_ _(top)_ and _Ankylosaurus._

But when we come to consider the distance of galaxies and quasars we must speak of thousands and millions of light-years. Because stellar distances are actually established by the measurement of tiny angles, astronomers frequently use another unit called the **parsec.** From a distance of one parsec, one astronomical unit subtends the angle of one second of arc. One parsec equals 3·2616 light-years, and Proxima Centauri is about 1·3 parsecs (1·3pc) from Earth. In talking of galactic and intergalactic distances, **kiloparsecs** (kpc), **megaparsecs** (mpc), meaning 1000 and 1 000 000 parsecs respectively, may then be used.

In looking out deep into space we must not forget that we do not see objects as they are now, but rather as they were, there and then. We look out not just into the far distance, but also into the ancient past.

Opposite top right: **we see the Small Magellanic Cloud as it was about 200 000 years ago, when woolly mammoths still roamed the Earth** *(opposite top left).*

Below: **the Local Group may form part of a great supercluster, which is centred on the Virgo cluster 40 million light-years away.**

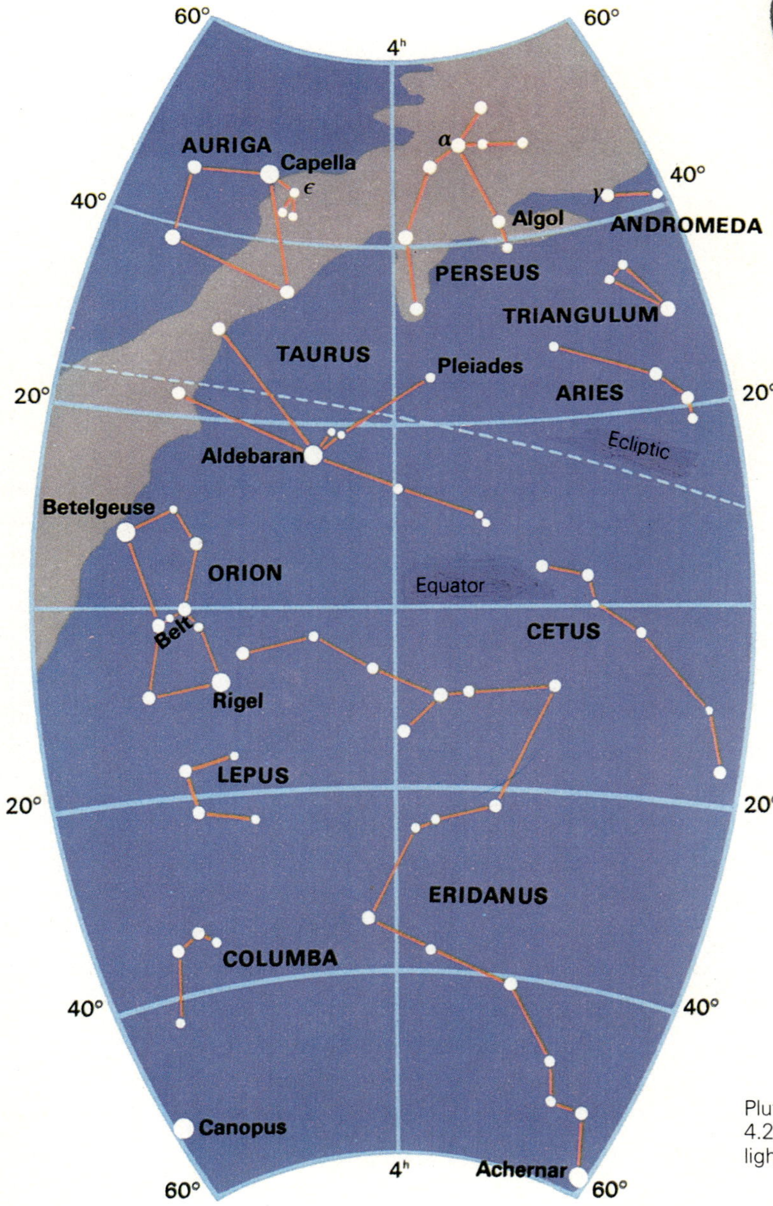

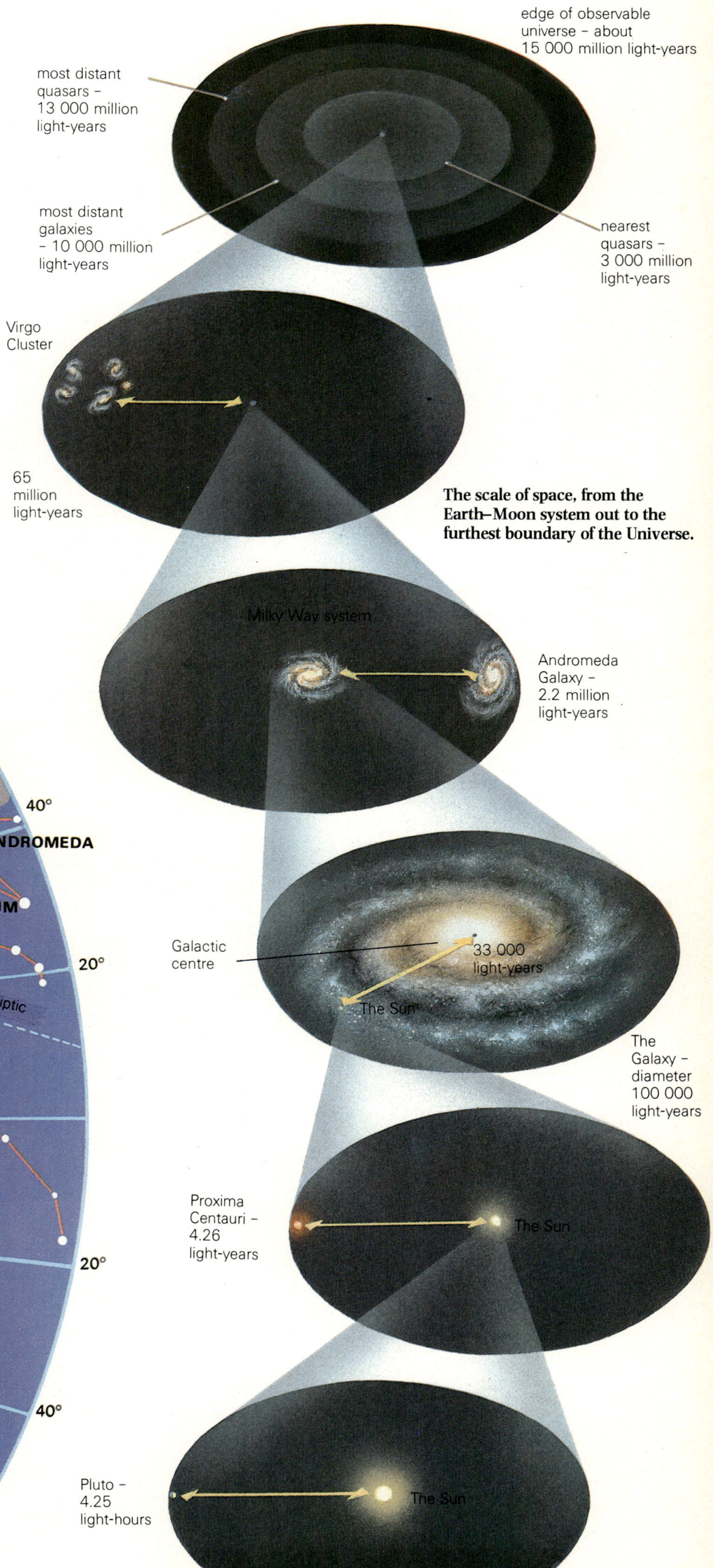

The scale of space, from the Earth–Moon system out to the furthest boundary of the Universe.

The Solar System

Once the motions of the planets had been understood (see pages 10-11), and with the advances in telescope design, a true idea of the scale of the Solar System could at last be formed. If the distance of just one planet (e.g. Earth) could be established all the others would follow from Kepler's Laws. The most accurate modern value for the all-important distance of the Earth from the Sun is just under 150 million kilometres (149 597 870km to be exact), and this has the special name of the astronomical unit. The major planets are spread out in space from Mercury at 0·39 astronomical unit all the way to Pluto at an average distance of 39·4 astronomical units – so far that it takes light, travelling at nearly 300 000km per second, about $5\frac{1}{2}$ hours to reach Pluto from the Sun. (It takes about 8 minutes to reach the Earth.)

The four **inner planets** – Mercury, Venus, Earth and Mars – have many similarities and

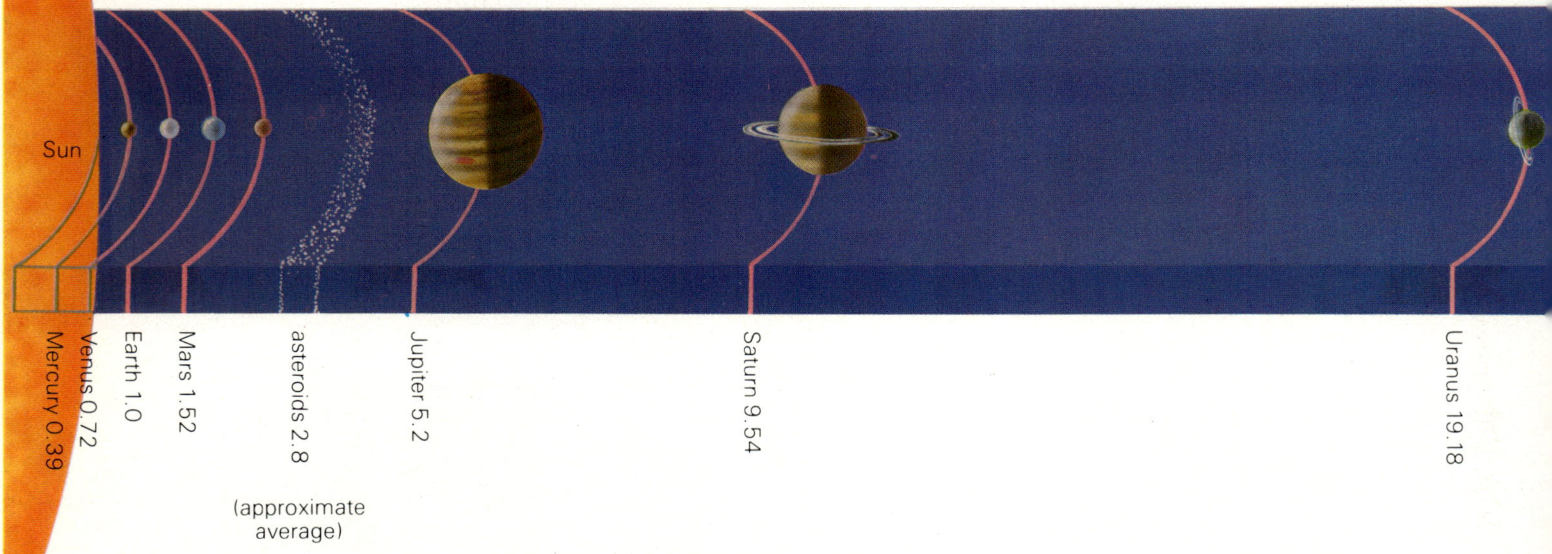

The formation of the Solar System from a cloud of dust and gas took place about 4·6 thousand million years ago, and included much material from older stars.

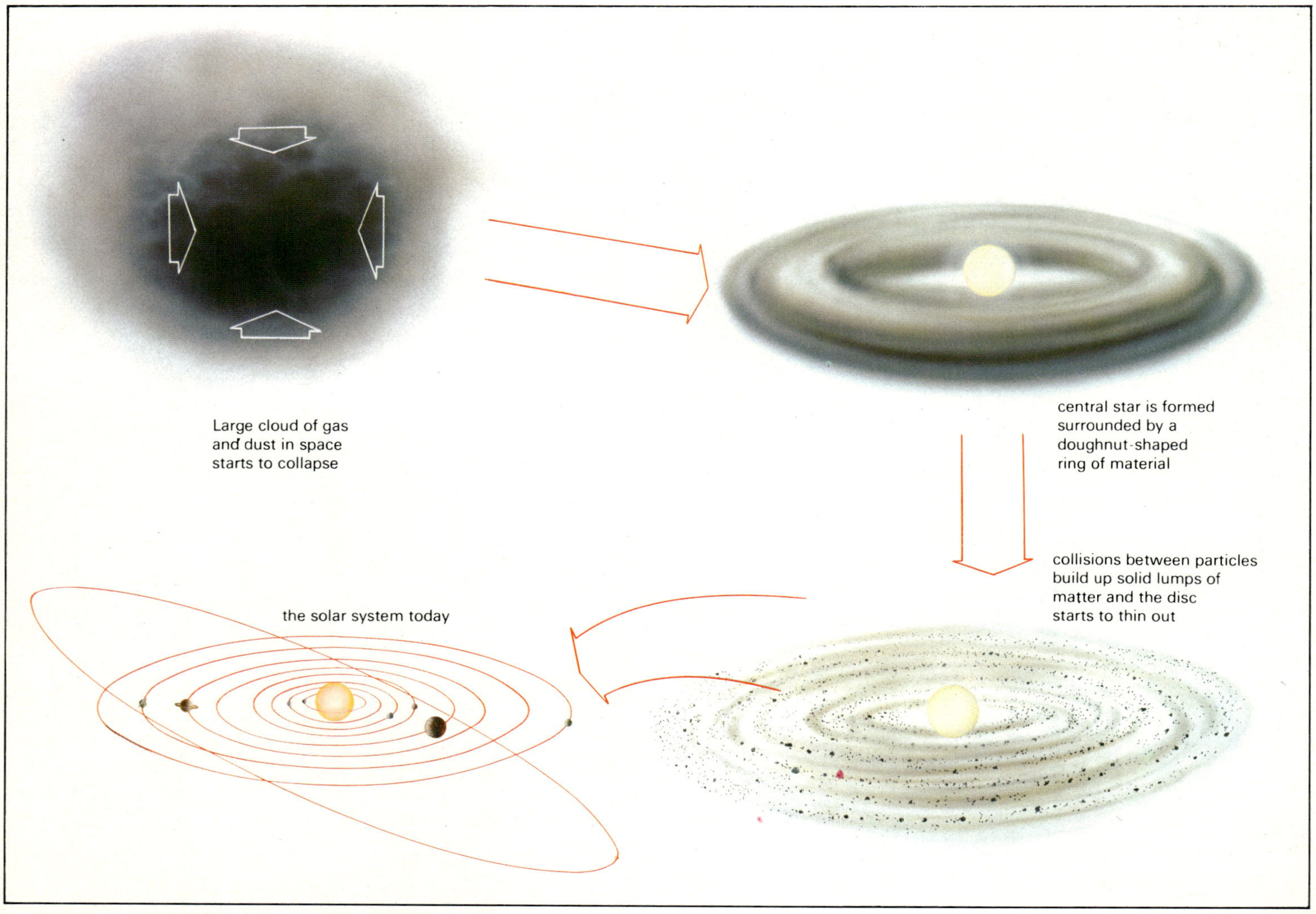

are fairly dense rocky bodies. They are often collectively described as the **terrestrial planets.** Four more – Jupiter, Saturn, Uranus and Neptune – are frequently called **gas giants,** as, although they are large and massive, they are not nearly so dense as the inner planets, and do in fact consist mainly of various gases. Distant Pluto is difficult to observe and study, but it is doubtless mainly composed of ices, as are many of the satellites accompanying planets in the outer Solar System.

Only Mercury and Venus have no moons at all. The two most massive planets, Jupiter and Saturn, have very extensive satellite families. There also exist numerous smaller bodies in orbit around the Sun – both the generally rocky minor planets, most of which are between the orbits of Mars and Jupiter, and the icy comets, which may swing out towards the nearest stars.

It is now generally accepted that, apart from hydrogen and helium, all the elements found in stars, planets, and even our bodies, were formed within the interiors of earlier stars, and have been dispersed into space in various ways. Where gas and dust are gathered together in the enormous clouds within our Galaxy, new stars are forming all the time. From one of these clouds the Sun was born about 4·6 thousand million years ago, belonging as it does to the second, or possibly even the third, generation of stars.

It is not known precisely how the planets were formed, although very many theories have been proposed in the past. It is believed that they most probably arose at about the same time as the Sun itself, and out of similar materials, rather than that they were once part of the Sun. Many questions remain, however, which only future investigations can hope to answer.

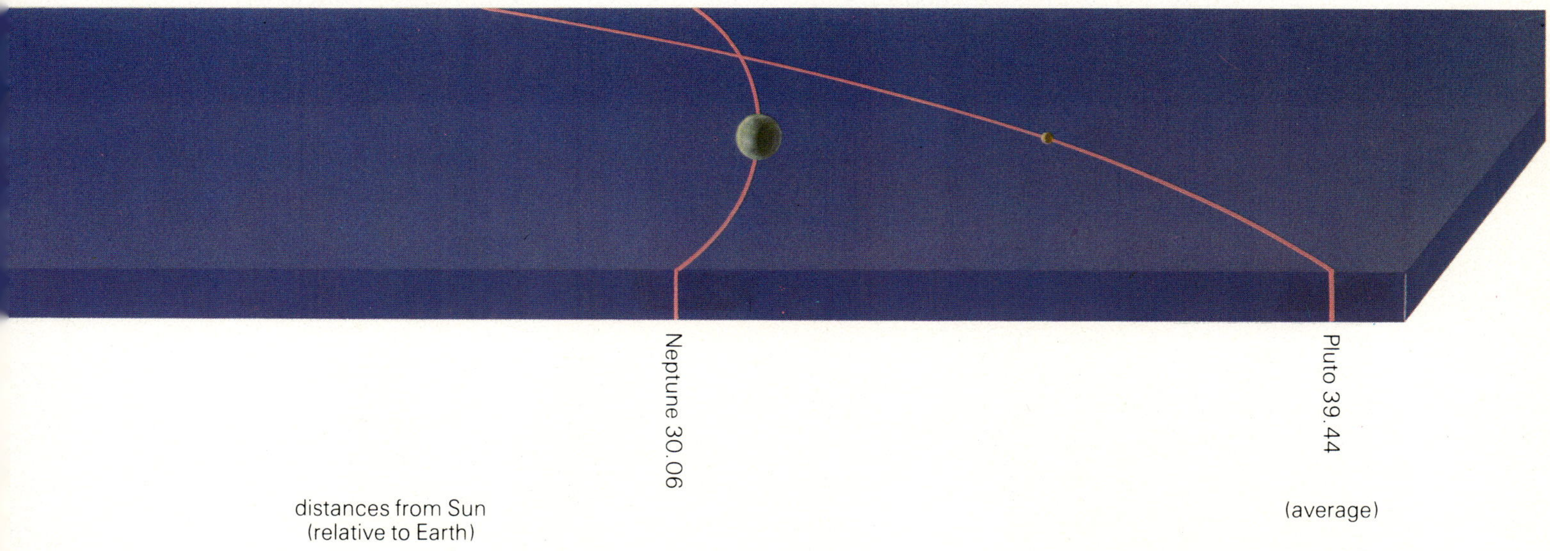

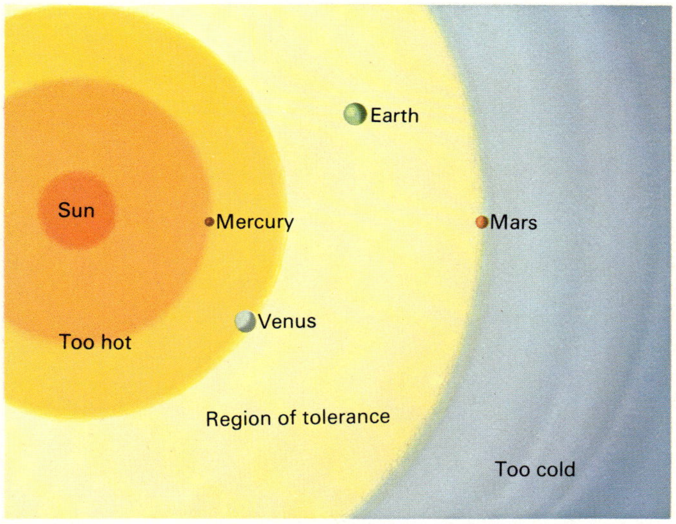

Below right: **an old theory of the origin of planets suggested that a passing star drew material from the Sun.**

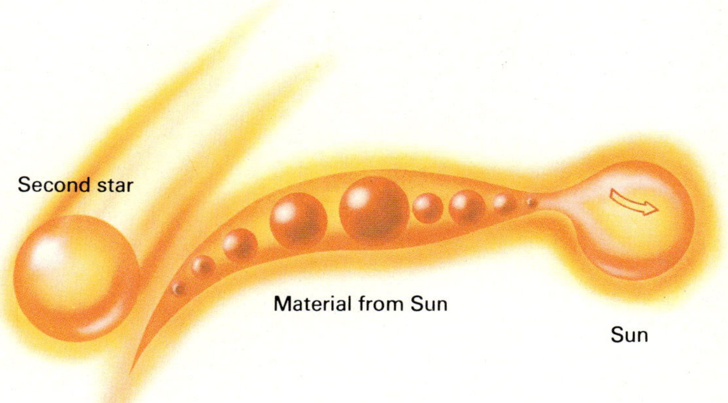

Opposite top and top: **the relative distances of the planets are shown in this diagram. However, the planetary sizes are not to scale.**

Above left: **conditions for life to exist are only found within a very narrow zone in the Solar System. Venus could once have been suitable but it is now too hot, and Mars is too cold.**

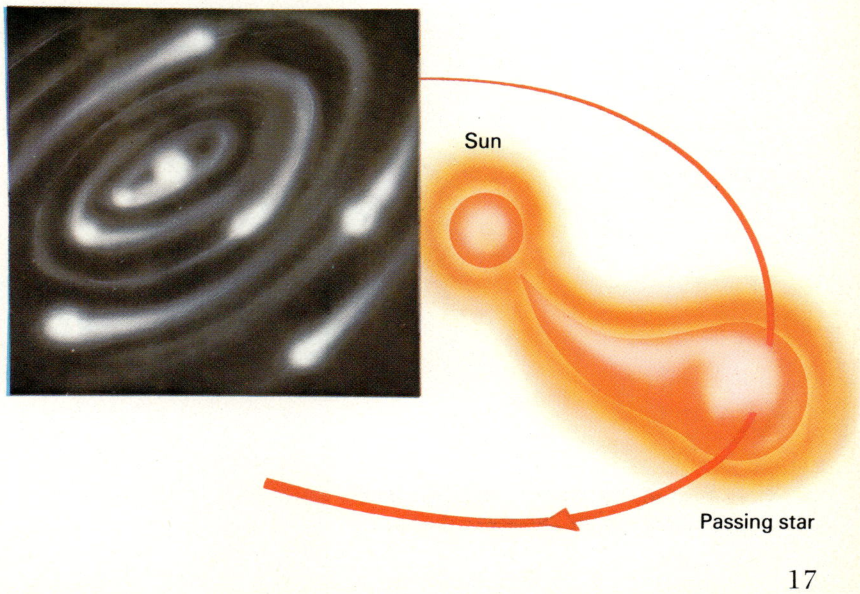

Right: **the planets probably formed from matter like the Sun's, although one theory requires the material to have come from a passing star.**

The Sun

The Sun is so important to us that it is sometimes difficult to think of it as a star, and a quite ordinary one at that. Although (as we shall see) some stars are hundreds of times larger, the Sun is certainly big. It is about 1 392 500km across, or nearly 110 times the size of the Earth. This whole vast globe consists of gas at a very high temperature, i.e. about 6000°C at the visible surface, rising to over 15 million °C in the core where hydrogen is being 'burnt' to helium with a tremendous release of energy. Although consuming about 4 million tonnes per second, the Sun has a calculated lifetime of about 10 thousand million years, and it is about half that age at present. When special equipment is used the Sun may be examined in detail. *Never* look at the Sun with binoculars or any form of telescope, or even stare at it with the naked eye. Its brilliance and heat are so great, even when low on the horizon, that damage or blindness will result: *it is not worth the risk.*

The most noticeable features of the surface

Energy is produced by nuclear reactions at the centre of the Sun, but surprisingly, it takes about 10 million years to reach the surface.

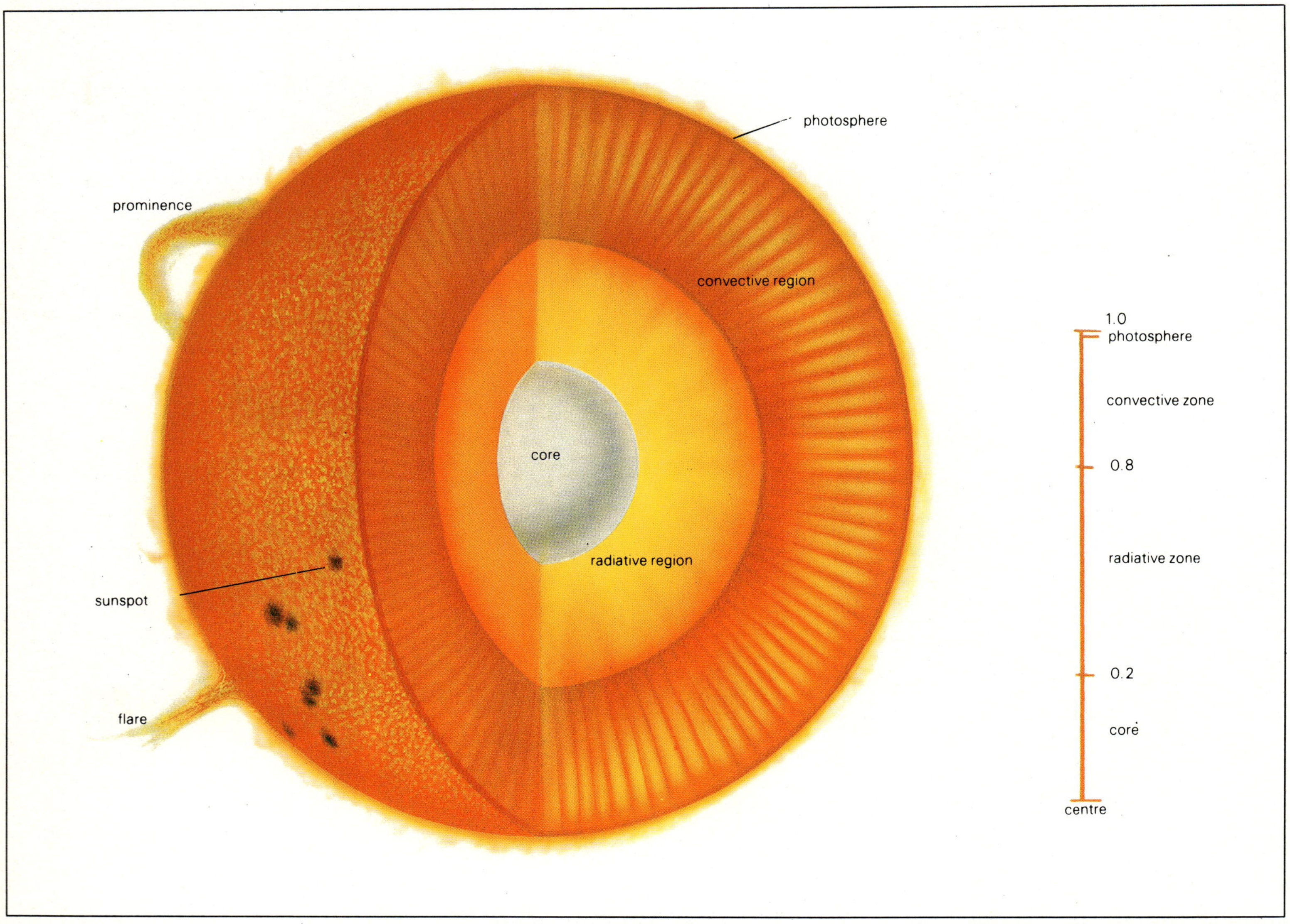

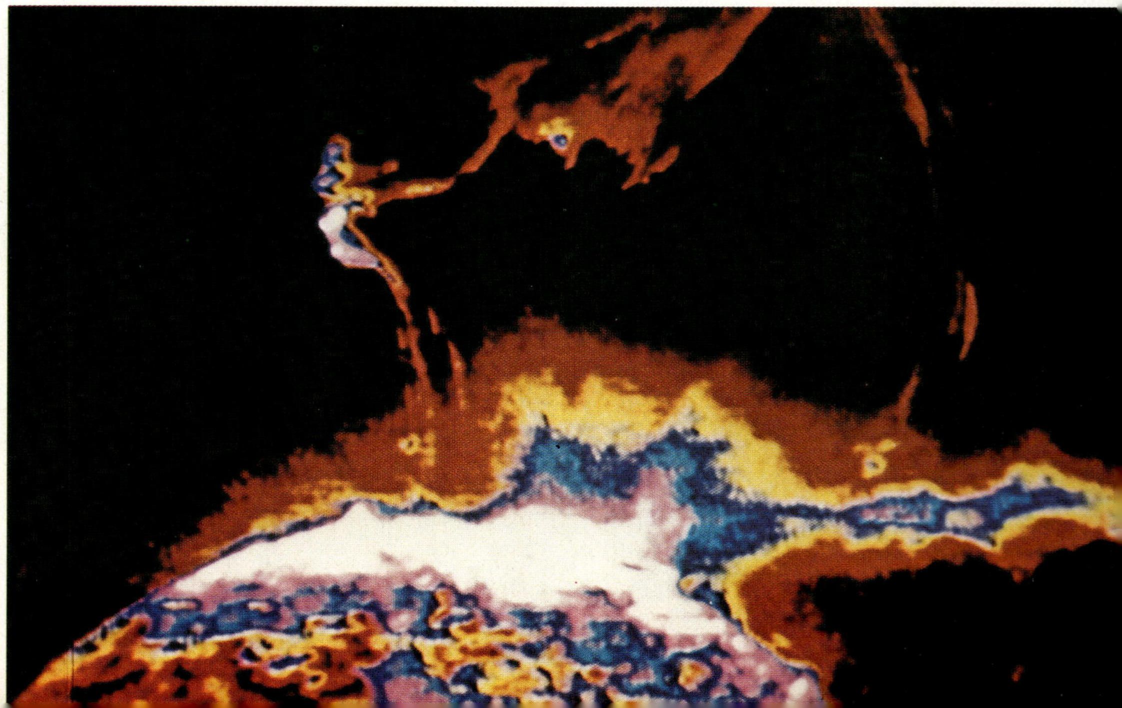

A 'false-colour' image in ultraviolet light of an active prominence on the surface of the Sun.

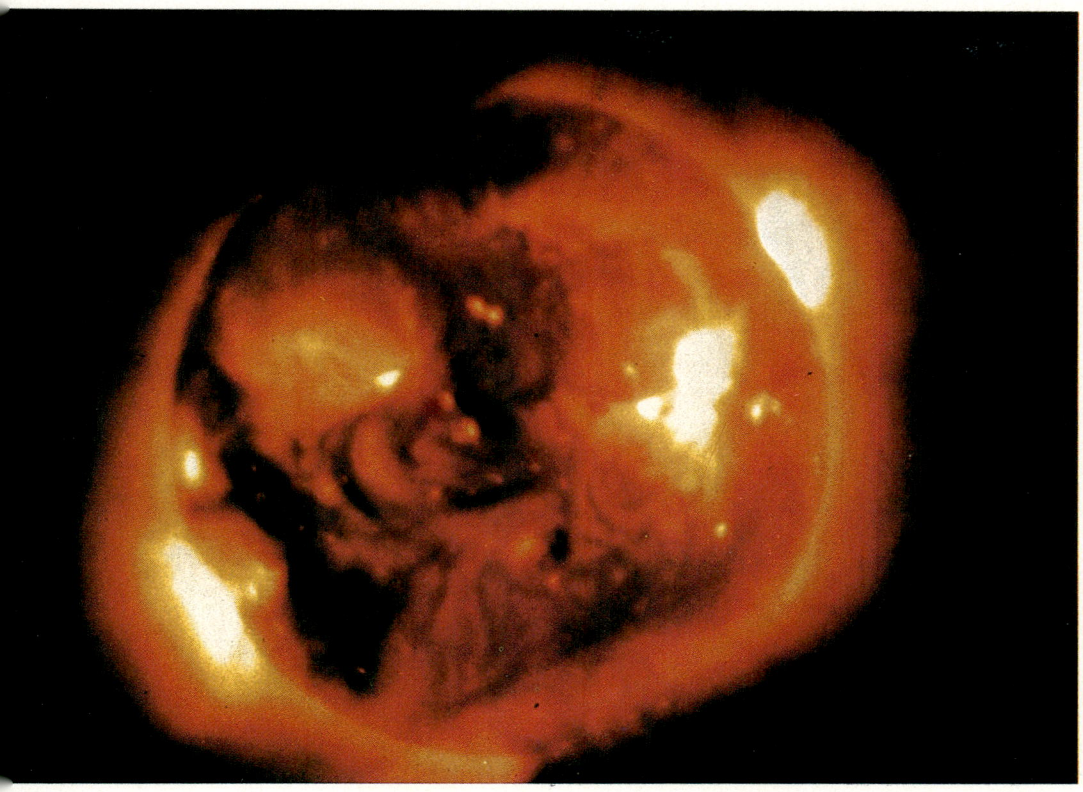

Left: **bright flares and dark 'coronal holes' are seen in this X-ray picture of the Sun obtained from Skylab.**

Above: **by using light in a very restricted range of wavelengths it is possible to map the distribution of temperatures and activity upon the surface of the Sun.**

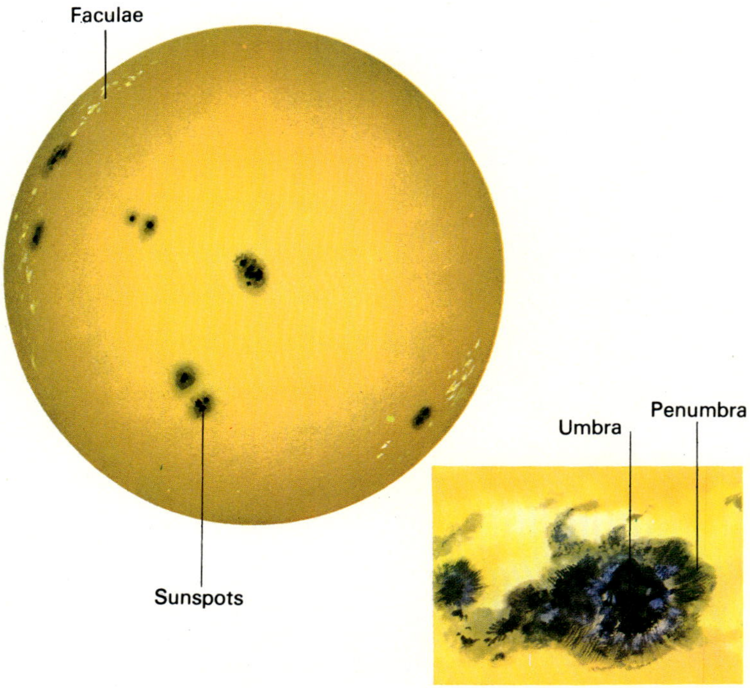

Sunspots often occur in pairs, or groups, linked by strong magnetic fields, unlike the bright, solitary faculae.

Below: **many satellites such as this Orbiting Solar Observatory– OSO 8– and various interplanetary probes keep a continuous watch on the Sun's activity.**

are **sunspots,** areas which appear dark by contrast with their brilliant surroundings, but which are still very hot (4500°C). Some of these may be larger in size than the Earth, and they often occur in groups, especially pairs. Individual spots may last for a few days or weeks, but the centre of activity gradually migrates towards the equator, before starting anew at high latitudes. The overall numbers fluctuate with a period of roughly 11 years and this sunspot cycle has often been proposed as causing changes in the weather on the Earth.

It is also frequently possible to see **prominences** around the edge of the Sun – vast loops and flame-like structures, sometimes extending outwards for hundreds of thousands of kilometres. Although they look as if they have been shot out from the Sun, most frequently the material is actually streaming *down* from the tenuous outer 'atmosphere' known as the **corona.** This stretches far beyond the main solar disc, and, except during solar eclipses, can only be seen with specialized equipment or from spacecraft. Although the number of atoms in the corona is very low, they are moving so fast that its temperature is about one million °C.

The Sun has a strong magnetic field which not only controls the position of sunspots on the surface and of the plumes and rays of the corona, but extends far out into space. It guides the stream of charged particles known as the **solar wind** which is constantly flowing from the Sun. It similarly channels the motion of those particles ejected from violent **solar flares,** and these frequently cause aurorae and radio interference when they reach the Earth and interact with its magnetic field.

Mercury

Sometimes, in the evening or morning twilight, it is possible to catch a glimpse of the innermost planet, Mercury. But this planet remains so close to the Sun that there are many modern astronomers who have never seen it. Ancient astronomers such as the Babylonians did notice it, and observed the rapid changes in position as it appeared to flit from one side of the Sun to the other. (Its complete orbit around the Sun only takes 88 days.) The planet's movement is fast, and it is not surprising that it was finally named after the Roman messenger of the gods.

The planet is faint and small (only 4878km across), and has always been so difficult to see that observers have never made out much detail on the surface, even with excellent telescopes. For many years the period of rotation on its axis was wrongly thought to be the same as its orbital period around the Sun; but radar and, later, spacecraft observations showed it to be just under 59 days. There was some excuse for the observers though, as it happens that Mercury always has the same face turned towards the Earth when it is best placed for observation.

Being so close to the Sun (on average just over one-third of the Earth's distance), Mercury becomes very hot and the temperature of certain parts of the surface can reach more than 425°C – well over the melting point of lead. On the side away from the Sun the temperature drops to around −175°C, so rocks probably erode by alternate expansion and contraction.

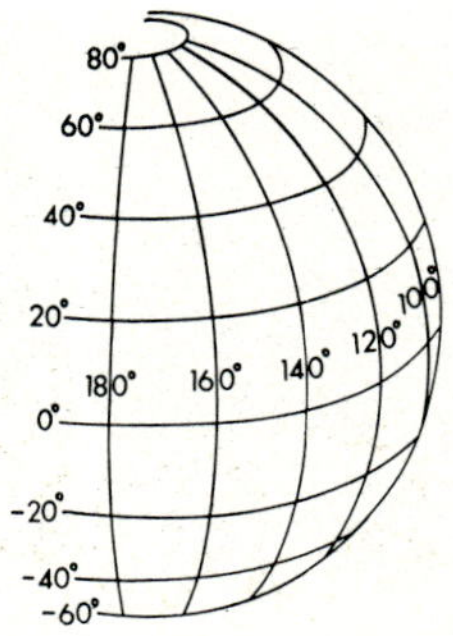

Below left: **the long scarps on the surface of Mercury may be wrinkles, formed when the planet cooled and shrank.**

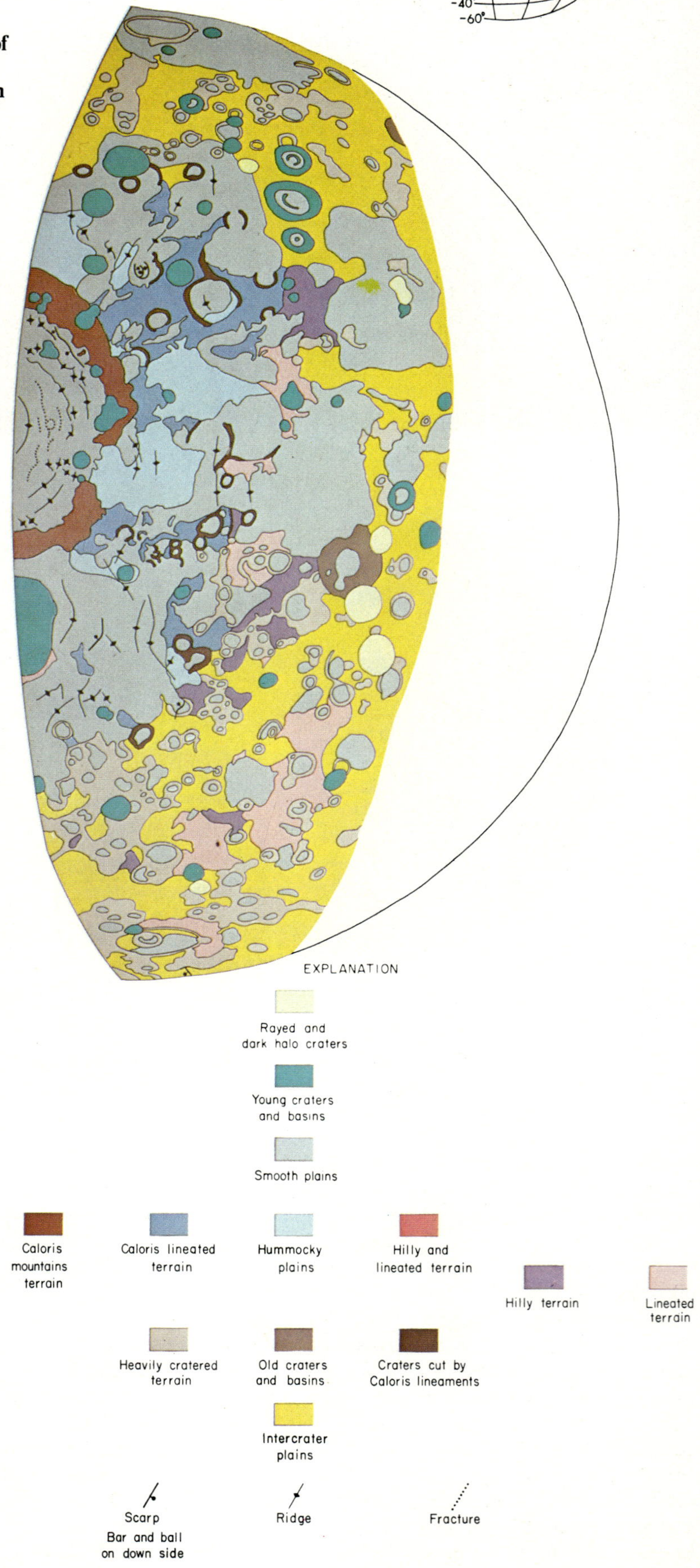

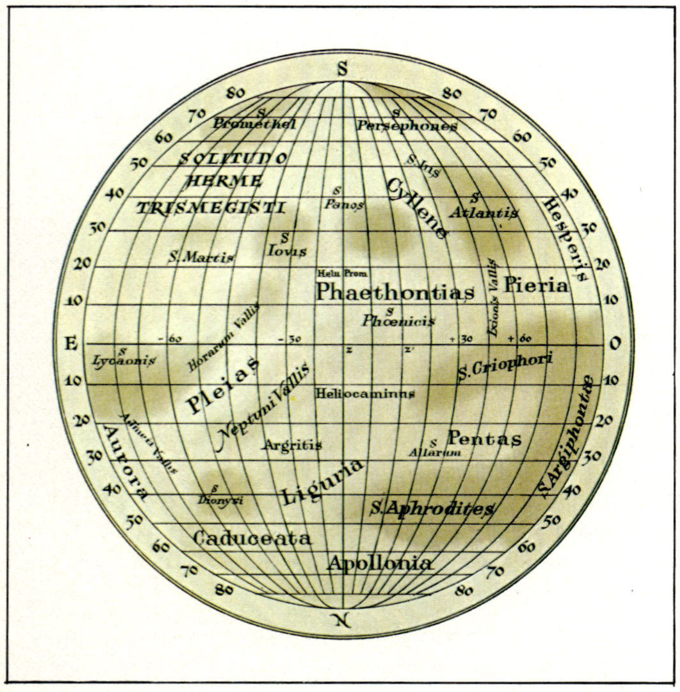

Below: **a geological map of Mercury made from the observations by Mariner 10.**

Opposite bottom: **prior to the spacecraft's visits to Mercury, this map by Antoniadi was the best that could be made from Earth.**

Right: **Mariner 10, the only spacecraft to survey Mercury, did so on three occasions in 1974 and 1975.**

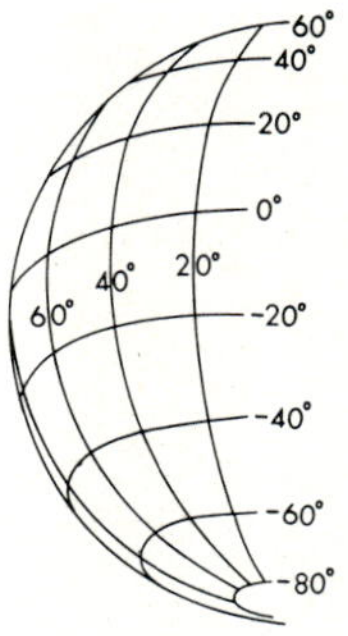

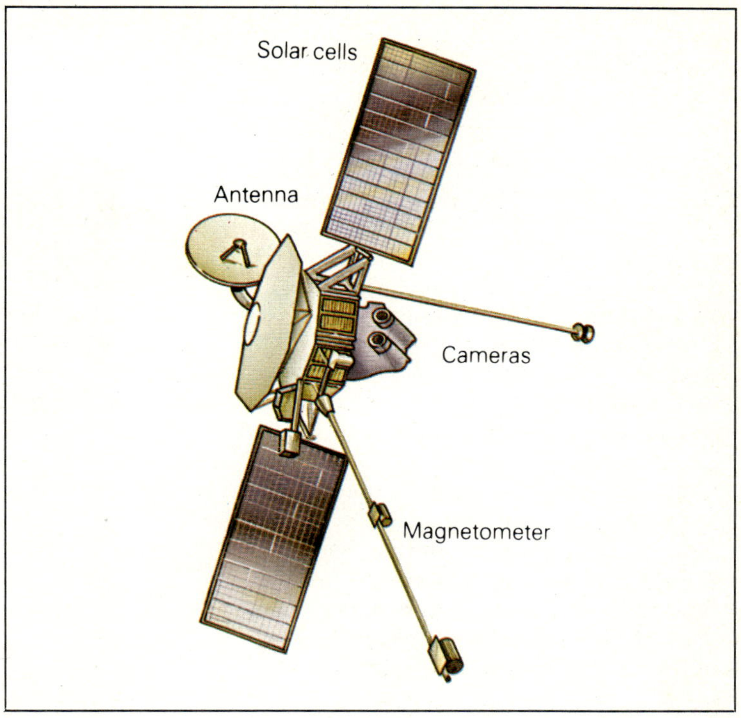

The orbit is very eccentric, and the planet's distance from the Sun varies by as much as 20 per cent. Small alterations in this very elliptical orbit could not be explained from Newton's gravitational laws, but were one of the earliest proofs that Einstein's theory of gravity, his Relativity Theory, was correct.

Only one spacecraft (Mariner 10) has visited Mercury, but this made three passes of the planet and sent back detailed pictures of roughly half of the surface. The general appearance is rather like that of the Moon, as there are some areas which are densely cratered and others which consist of flatter plains. Many of the craters show central peaks, but these are not as high, relatively, as on the Moon, because of Mercury's greater surface gravity. There is at least one large impact feature, the Caloris Basin, which is about 1500km in diameter. Running across the surface there are long scarps, which are not really similar to anything found on any other body. Mercury is very dense however, and is also magnetic, which suggests that it has a large core of iron and nickel. The scarps may have been produced when this core cooled after the planet's formation, and shrank, wrinkling the overlying rocky crust.

Above: **the interior of Mercury probably consists of a very large metallic core surrounded by a rocky outer layer.**

Venus

Although Venus comes closer to the Earth than any other planet, it is so heavily cloud-covered, with only occasional vague markings becoming visible, that practically nothing was known in detail until about 20 years ago. Its orbital period (225 days), its diameter (12 100km) and its mass (about 80 per cent of Earth's) were known, and that was about all. Even its day length was a mystery. But now, thanks to spacecraft and probes which have landed on the surface, and to radar measurements, we have much more information about the planet.

The rotational period is very long (243 days) and rotation is **retrograde** (clockwise as seen from over the north pole). This differs from the rotation of all the other planets, and also from the motion of the planets (including Venus itself) around the Sun. There is no satisfactory explanation of why this should be so, nor of why the 'day' is longer than the 'year'.

The surface temperature has proved to be 500°C– very hot indeed, and the same as, or slightly hotter than, Mercury. This heating is caused by the very dense atmosphere, which consists mainly of carbon dioxide, and which has a surface pressure more than 90 times higher than the Earth's. This produces a very

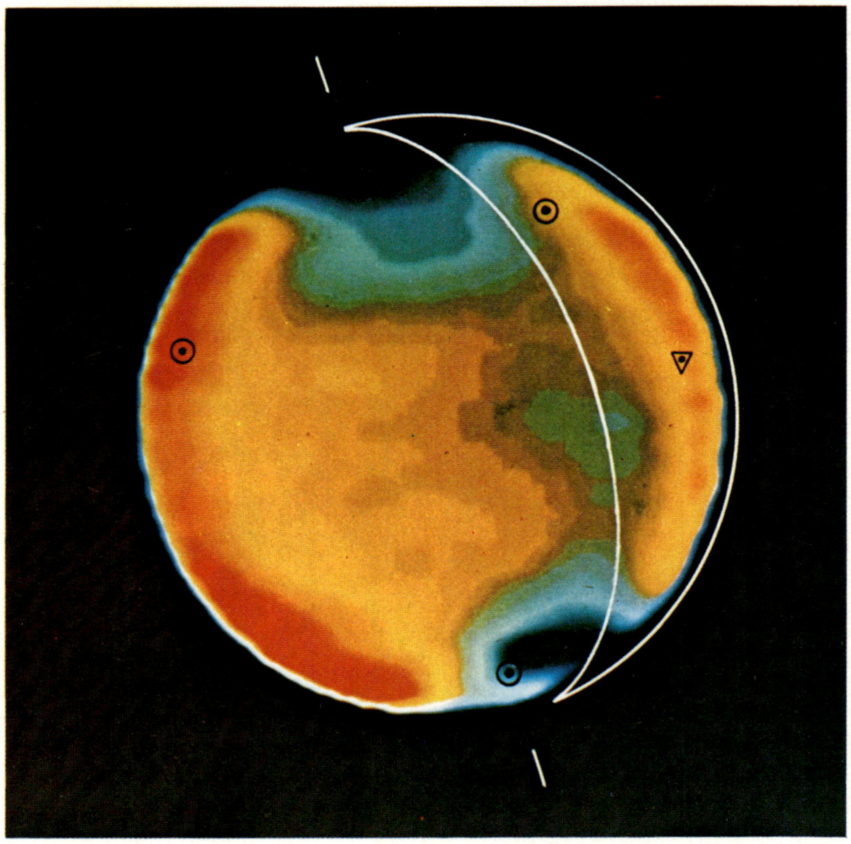

Above right: **the landing sites of the Pioneer Venus probes are shown on this colour-coded temperature map of Venus.**

Below: **this radar map shows the surface of Venus to be mainly flat, with a few higher regions, which are probably volcanic.**

Opposite top: **a recent Venera lander sent back this picture from the searingly hot surface of Venus.**

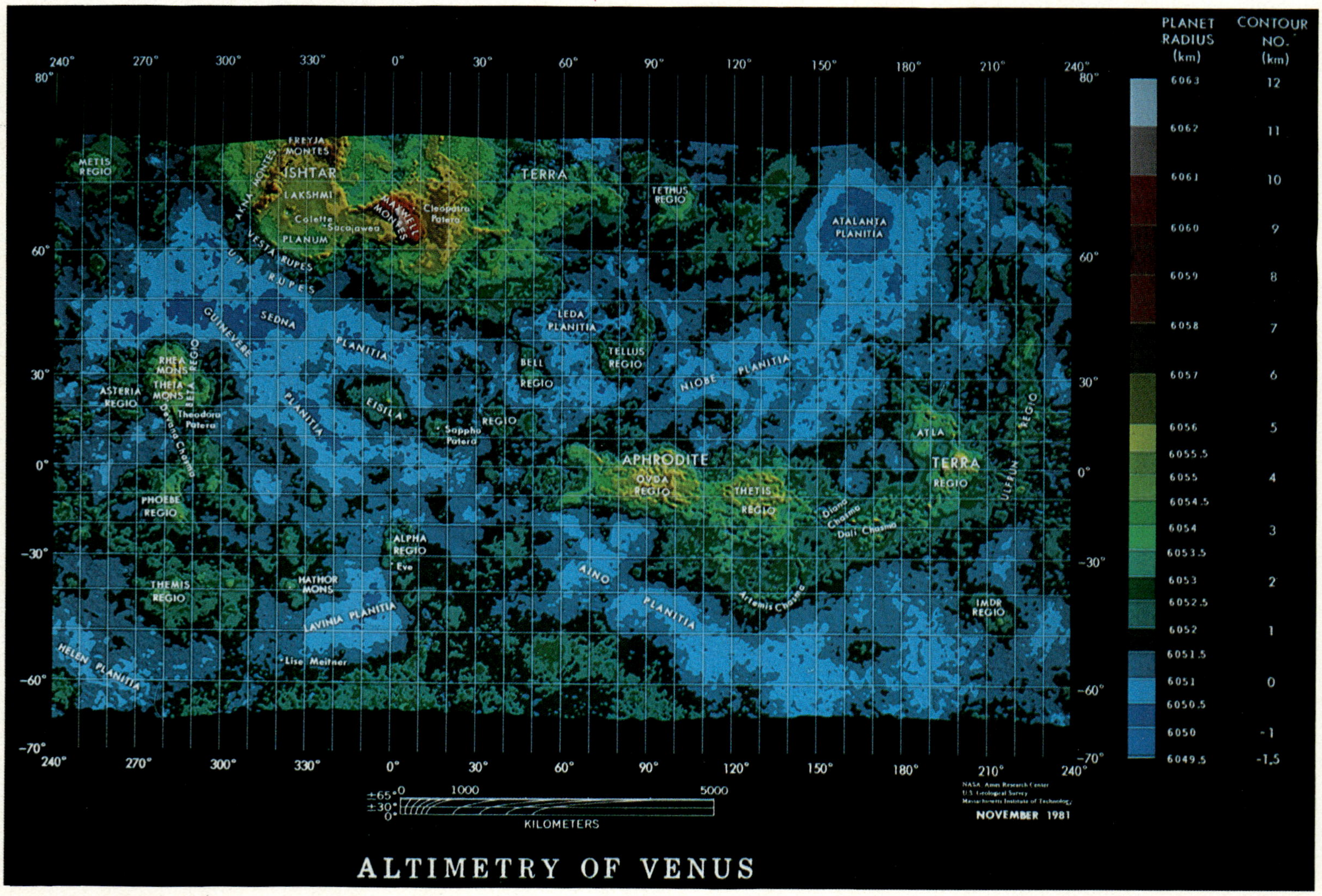

strong 'greenhouse effect': the atmosphere allows heat from the Sun to enter, but prevents the planet's heat from being radiated away to space. The high temperature also causes chemical reactions at the surface, and these have made the greenhouse effect even more extreme.

The dense clouds appear to consist of sulphuric acid droplets and are very high, lying about 45-65km above the surface. The topmost clouds race around the planet in four days, but lower down they spiral slowly towards the poles.

Measurements carried out by radar, both from Earth and from the Pioneer Venus Orbiter, show that Venus is generally fairly flat, but that there are a few high regions including one, Maxwell Montes, which rises to about 12km (about the same as the height of Mount Everest above the average level of the sea floor – rather than sea level – on Earth). Lightning has been detected here and over only two other high regions, and is thought to indicate currently active volcanoes.

The Pioneer Venus probes mainly provided atmospheric information, but the Venera landers have returned pictures of the surface, as well as analyzing the composition of the rocks. These appear to be mostly similar to the basalts of the sea floors on Earth, although there are indications of others resembling the lunar highlands and the terrestrial continents.

Before the greenhouse effect came into play, Venus may have had considerable oceans, and could have supported some form of life. As the water 'boiled away' it became the inhospitable planet we find today.

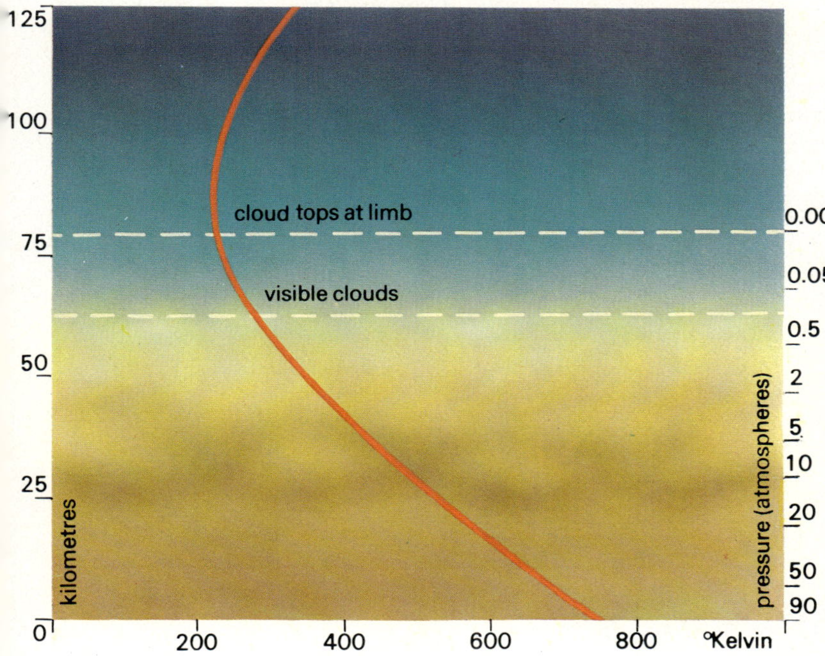

Above left: the very dense atmosphere of Venus has thick cloud layers high above the surface.

Left: One of the earlier Venera probes which parachuted down through the atmosphere.

Above right: ultraviolet light shows some markings in the uppermost layers of the clouds.

Right: an artist's impression of the deep rift valley on the surface of Venus.

The Earth

Many people find it difficult to think of Earth as a planet, but astronomers have always done so, and automatically use it as a standard of comparison. Now that astronauts and spacecraft have given us pictures of it from space, it is more generally appreciated how the Earth relates to other bodies in the Solar System.

What is different about the Earth as a planet? The most striking feature from space is the abundance of water, both in the oceans and in clouds. Quite unique, too, is the oxygen-rich atmosphere, so essential to animal life, which has mostly been created by the action of plants in converting carbon dioxide.

If all the water were removed, most of the surface of the Earth – the ocean floors – would appear as reasonably flat plains, quite like those on Venus. But then closer inspection would show the oceanic ridges and trenches, the mountain arcs, and all the other signs of current geological activity. Modern ideas of how all these features have been formed – through 'plate tectonics' and 'sea-floor spreading' – allow us to look at other planets and say with some confidence that none are so active as the Earth. Venus and Mars show some signs of activity, it is true, but their crusts are thick, and little has happened recently (that is, in the last few hundred million years). Io, the satellite of Jupiter, is very active, but its volcanoes are rather special and there is no comparison.

By the use of seismic waves we have been able to study the interior of the Earth, and we know that beneath the crust with its continents and sea-floors lies a thick rocky mantle, which contains most of the mass of the entire planet. Further within there is a liquid region, and then a solid inner core, both of which are probably metallic.

Right: **the Earth's magnetosphere is drawn by the solar wind into a long tail extending far out into space.**

Below right: **the Earth from space, showing how water dominates its appearance.**

Below: **the relative abundance of the various elements in the crustal layers of the Earth.**

Bottom: **aurorae occur when charged particles from the magnetosphere excite atoms in the upper atmosphere.**

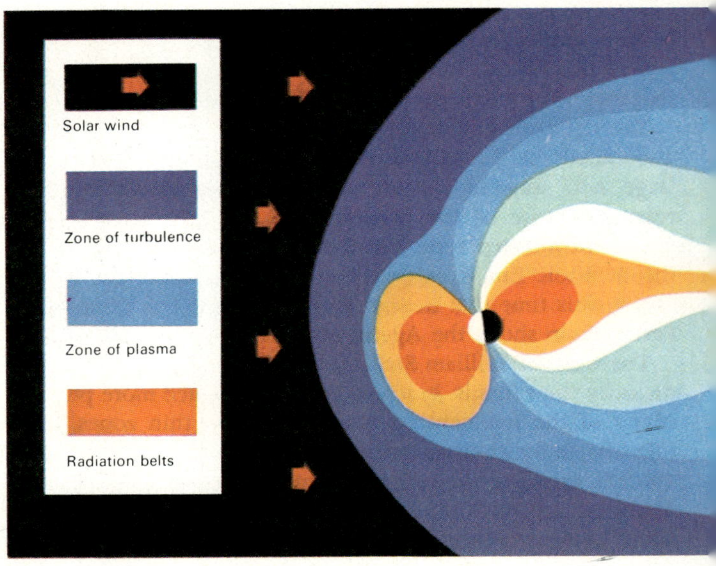

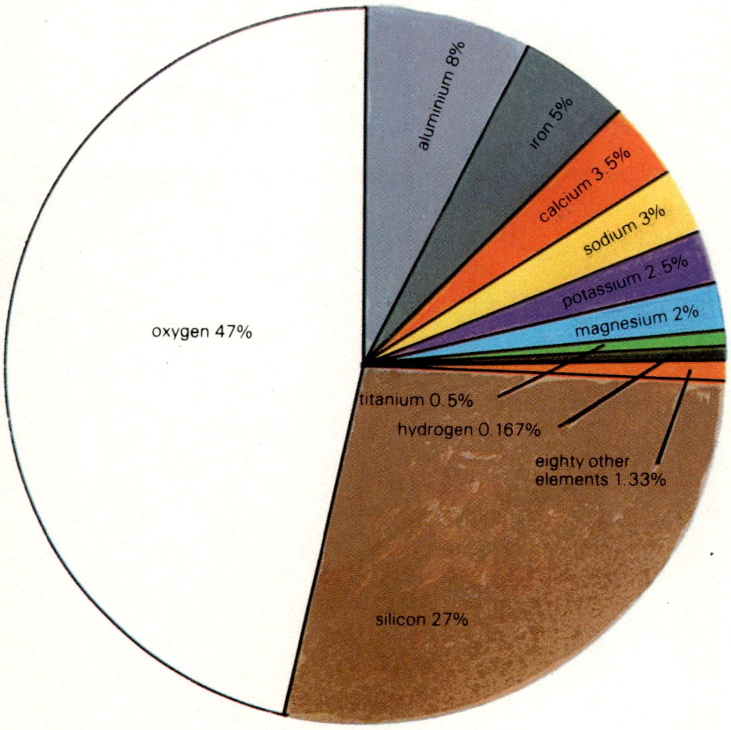

Right: **the tides are the most obvious effect of the Moon's gravitational attraction upon the Earth.**

The Earth's magnetic field is generated by the motion of the liquid outer core. Venus should have a similar interior, but its slow rotation does not produce general planetary magnetism. Jupiter and Saturn are fast rotators and both have strong magnetic fields, generated within their deep layers of liquid 'metallic' hydrogen. Each of these planets also have aurorae, caused by charged particles produced within their magnetosphere – their magnetic sphere of influence – cascading down into the upper atmosphere. On Earth the beautiful displays of the Northern and Southern Lights are closely linked with the number of sunspots and general activity upon the surface of the Sun.

Below: **the interior of the Earth as mapped by seismic waves from earthquakes and explosions.**

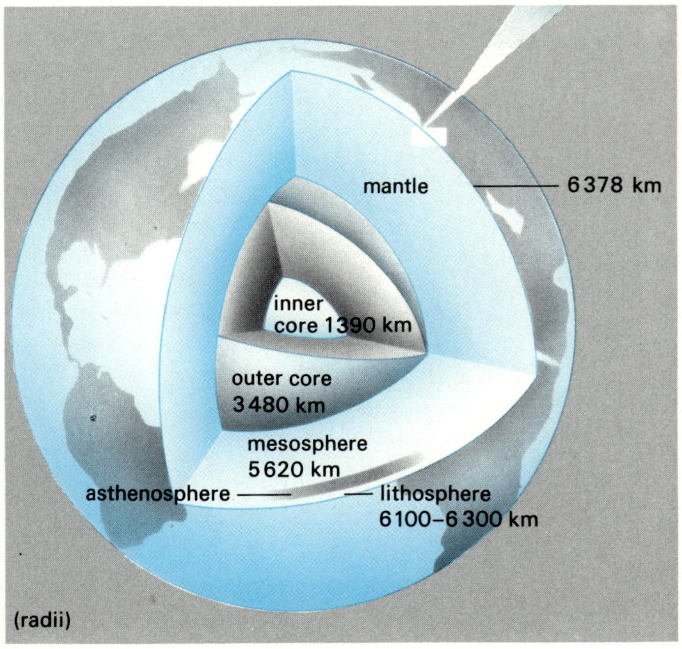

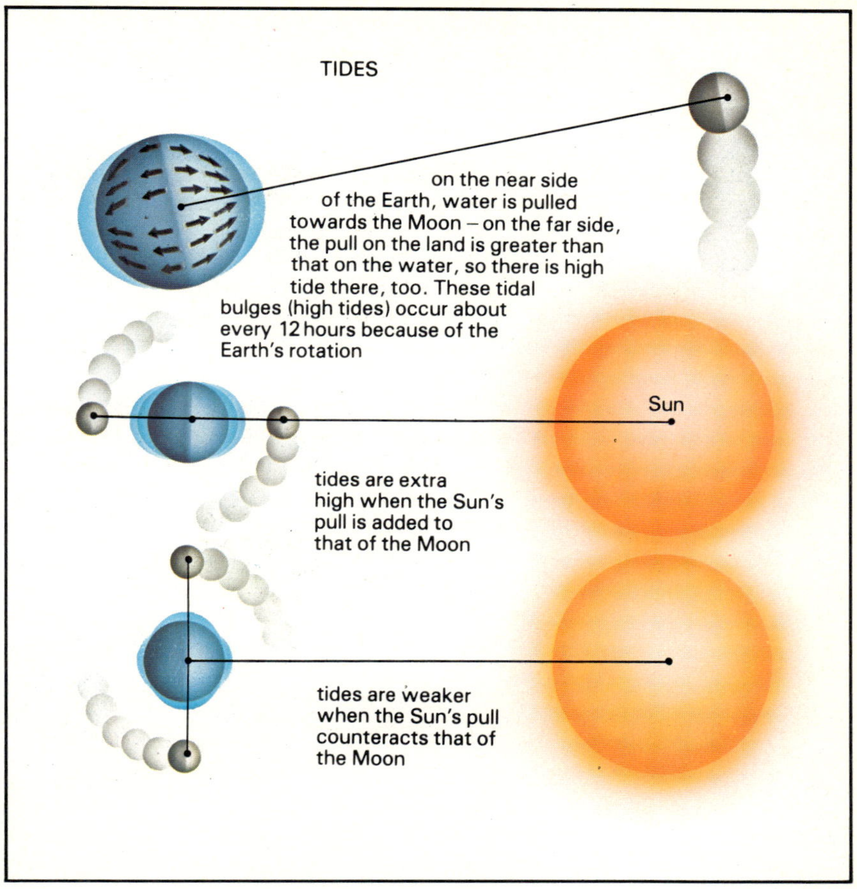

Below: **the matching of rocks across the Atlantic is evidence for continental drift on Earth.**

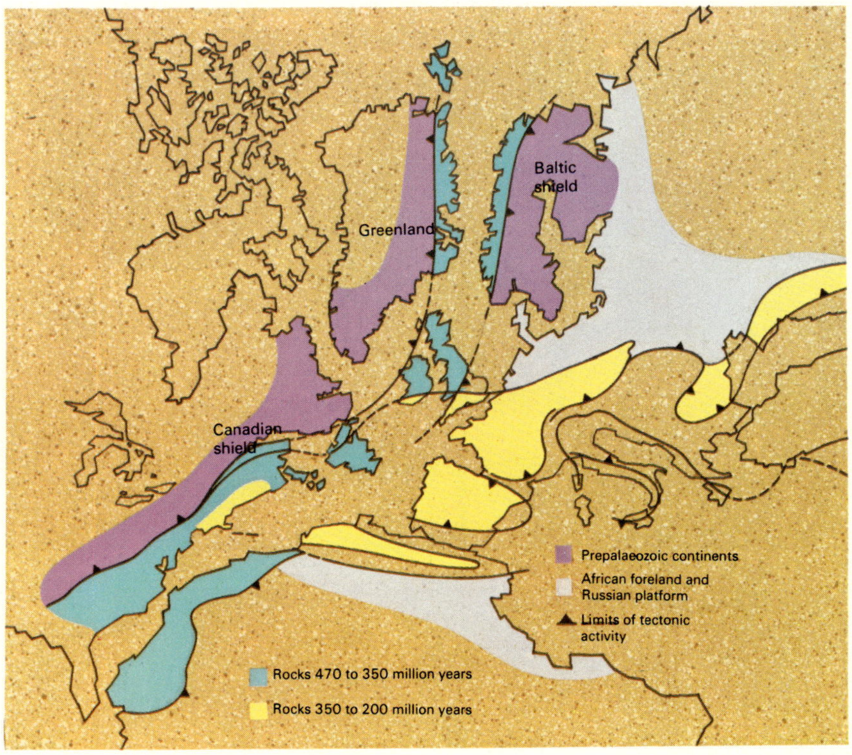

Below: **both the Moon and the Earth actually revolve about the centre of mass of the system, the barycentre, which lies beneath the surface of the Earth.**

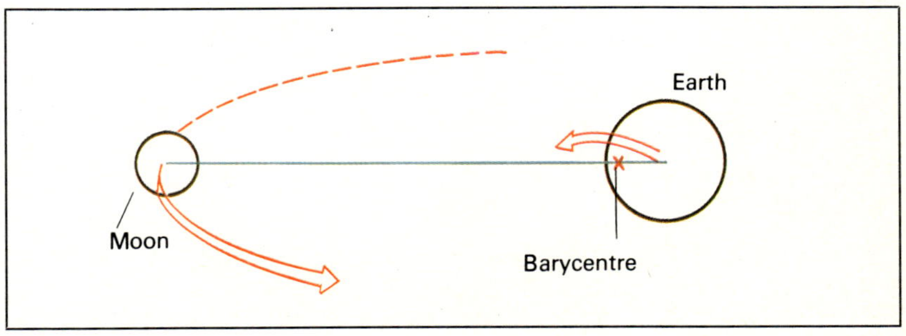

Eclipses of the Sun and Moon

In its monthly orbit the Moon occasionally passes directly between the Sun and the Earth, to give rise to a **solar eclipse,** or else moves into the Earth's shadow, and is itself obscured in a **lunar eclipse.** As the orbit is tilted at about 5° to the Earth's own orbit around the Sun, eclipses do not take place at every New and Full Moon, but only when the three bodies are in perfect alignment. Even so, eclipses are not particularly uncommon, and they occur as many as seven times in a year.

Whilst a lunar eclipse can be seen by everyone on the dark side of the Earth, the chance of seeing a total solar eclipse from any particular place comes about once in 300 years. This is a great pity, as a total solar eclipse is one of the most beautiful sights in nature, and many astronomers are prepared to travel round the world to see one.

When a solar eclipse takes place, the shadow of the Moon sweeps across the Earth in a narrow path no more than about 300km wide. Even within this track the Sun may not be completely hidden: the Moon's orbit around the Earth is elliptical, so that the apparent size of the Moon does vary. It may happen to be too small to cover the whole of the Sun's disc, which will then appear as a ring surrounding the Moon; this is called an **annular eclipse.** Outside the narrow track of the total or annular eclipse there is a much wider zone where a **partial eclipse** takes place, only a portion of the Sun being hidden. (Do please remember that viewing a partial or annular eclipse can be just as dangerous as attempting to look at the Sun, and that proper methods must always be used.)

A total eclipse is always spectacular, but no-one ever knows quite what to expect. As the Moon covers the last tiny portion of the Sun, bright pink prominences may come into view, while all around spread the plumes and rays of the pearly corona. Unfortunately the sight cannot last long – $7\frac{1}{2}$ minutes at the most– as the shadow races its way across the Earth. At the beginning and end of the total phase it may be possible to see the brilliant 'diamond ring' effect, or 'Baily's beads' like a necklace round the Moon, as sunlight strikes through valleys on the surface.

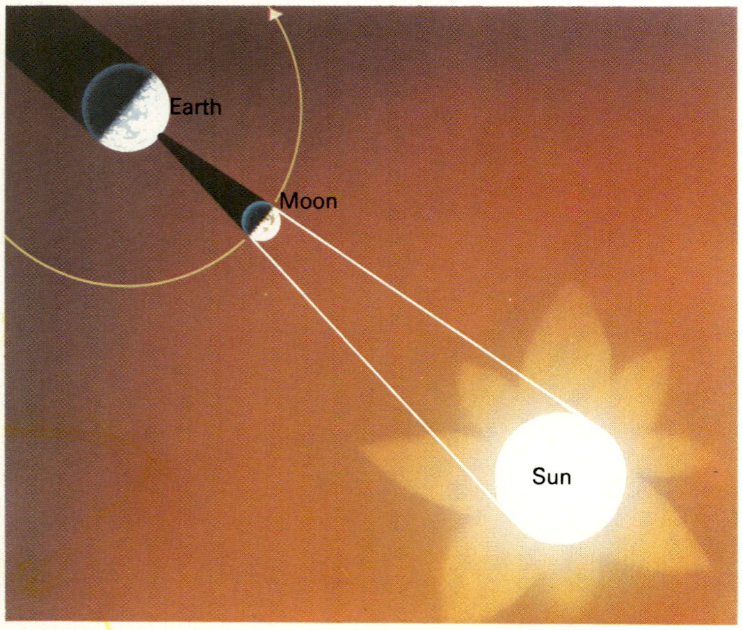

Opposite bottom left: **there are just a few occasions in a year when the Sun and Moon are aligned and a solar eclipse can occur.**

Opposite bottom right: **the unpredictability of eclipses gave rise to many strange ideas, such as the ancient Chinese belief that the Sun was devoured by a dragon.**

Opposite top: **the splendour of the corona can only be seen when the light from the surface of the Sun is blocked by the Moon.**

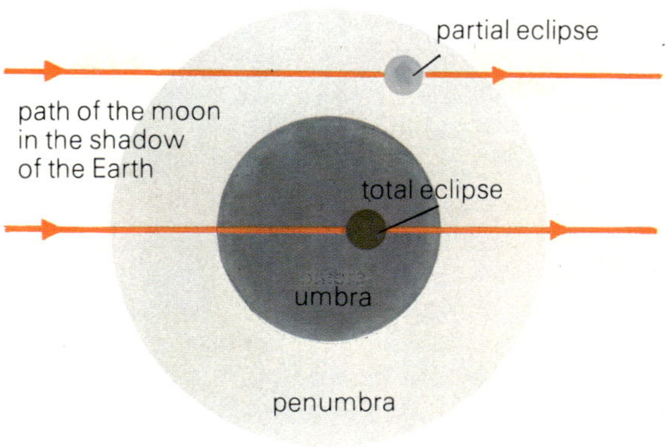

A lunar eclipse may not be so dramatic, but is still very striking. As the Earth's shadow is so much larger than that of the Moon, lunar eclipses last much longer, and the total phase can be nearly two hours. Because of the effect of the Earth's atmosphere, which refracts some of the sunlight, the shadow is very rarely completely dark. Usually the Moon appears a reddish or coppery colour, although on just a few occasions it has been said to disappear completely.

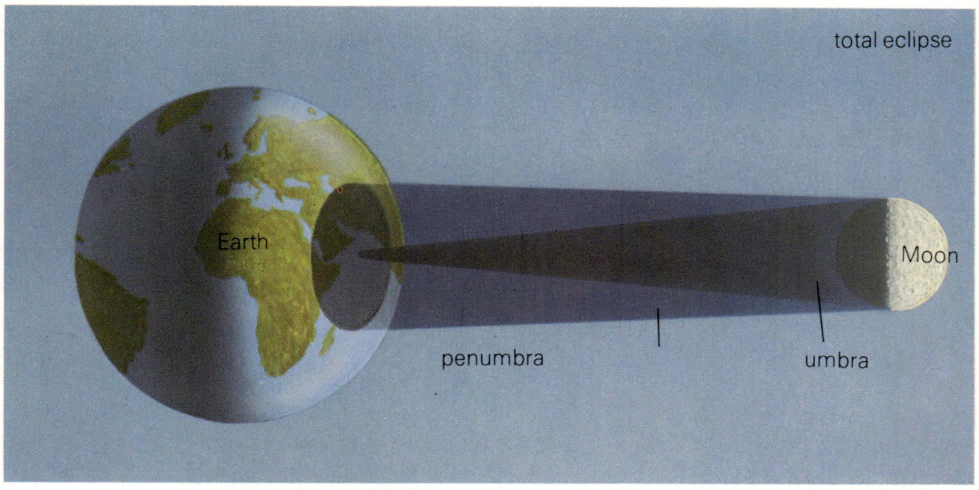

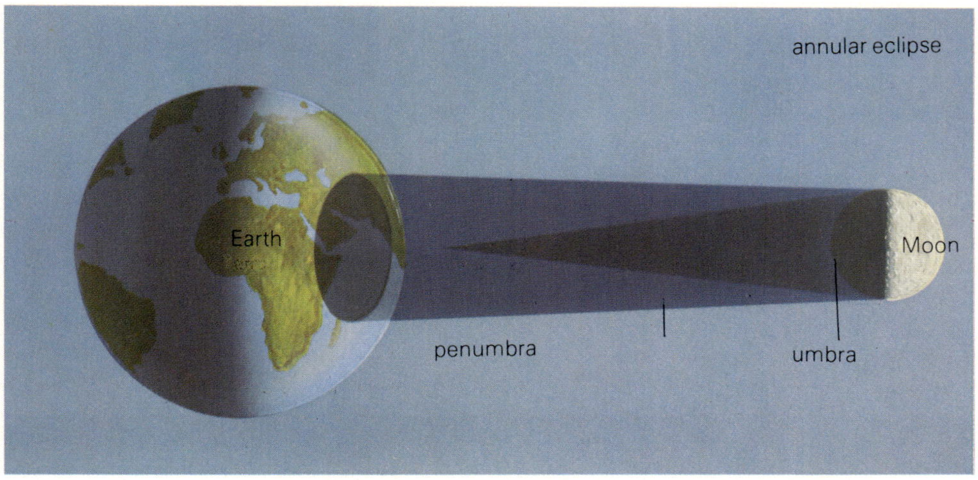

Above: **during a lunar eclipse only red light, refracted by the Earth's atmosphere, can illuminate the surface of the Moon.**

Left: **if the inner shadow of the Moon does not reach the Earth an annular eclipse is produced rather than a total one.**

Top: **the Moon sometimes passes through the Earth's outer shadow, but such a penumbral eclipse is usually too weak to be seen.**

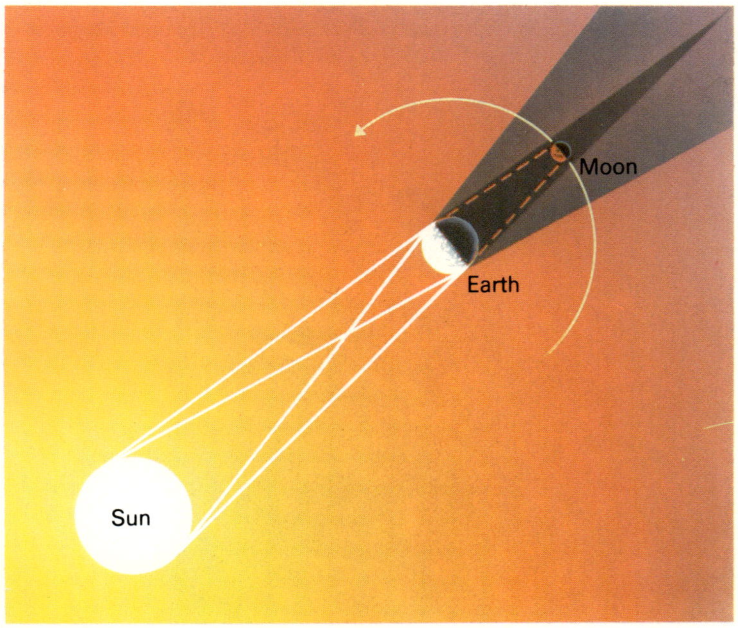

Left: **eclipses of the Moon can only occur when it is opposite the Sun in the sky, that is at Full Moon.**

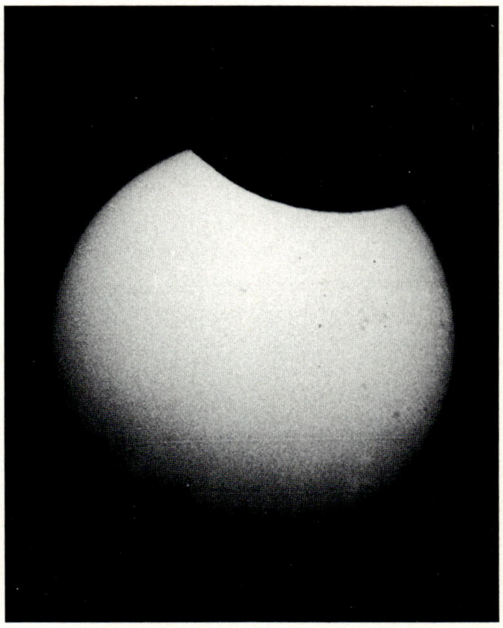

Although total eclipses are rarely seen from any place, partial eclipses are far more frequent.

The Moon

As a satellite the Moon is very large and massive, and only in the case of far distant Pluto and its satellite Charon do we find another system which, like the Earth and Moon, can reasonably be called a 'double planet'. Earth's ocean tides are the most obvious effect of the Moon's gravitational pull, although it does also cause tides in the ground itself, and in the atmosphere. Being so much more massive the Earth exerts even stronger control over the Moon, locking its rotation so that it now turns on its axis once every orbit, always keeping the same side towards us. However, although the rate of rotation is constant, the Moon's speed in its elliptical orbit is not, and it appears to swing backwards and forwards slightly on its axis. The result of this is that over a period of time we are able to see slightly more than half the surface (actually about 59 per cent).

Many names on the Moon, such as Mare Nubium (Sea of Clouds), or Oceanus Procellarum (Ocean of Storms), preserve memories of the time when it was thought that the dark areas were bodies of water. It has long been known that this is not the case, and that the Moon is an airless, waterless and lifeless world, but astronomers still use the Latin term 'maria' for the low volcanic plains, or 'seas'.

On the near side of the Moon the division of the surface into two types of area is very striking. On the one hand there are the dark, low-lying and comparatively featureless maria, and on the other the bright, densely-cratered highlands. We now know that the far side is mainly covered in craters and that there are few mare areas. For many years there was considerable argument about whether the craters had been formed by volcanic action, or whether they were results of impacts. It has now been established that the impact theory is correct, and that many of the large circular mare basins are enormous impact craters which have been filled with lava flows from inside the Moon. There are a few low domes, but nothing which could really be called a volcano.

Some of the well-defined craters are very large, and those of about 200km and more across are frequently known as walled plains. Small craters may be bowl-shaped, but the larger ones become more complex and may show terraced sides and often a central peak. It is thought that this particular feature is formed by the rebound of the underlying rocks after the explosive impact has excavated the initial crater. At Full Moon many craters show conspicuous ray systems formed by tiny fragments ejected by the explosion.

Below: **as the Moon orbits the Earth, it goes through its regular cycle of phases every month.**

Centre left: **the Moon spins once for every orbit so that the same face is always turned towards the Earth.**

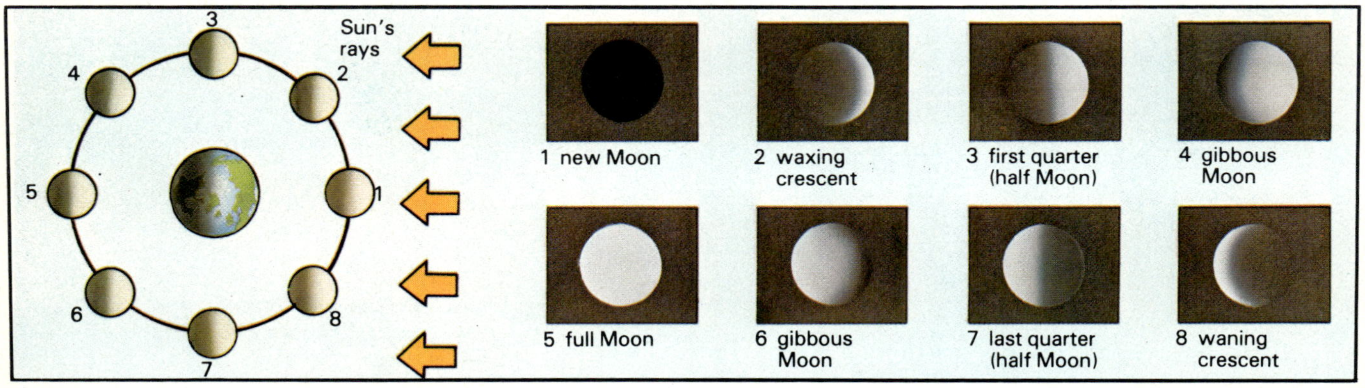

Far right: **Luna 13 was one of the early soft-landing lunar probes.**

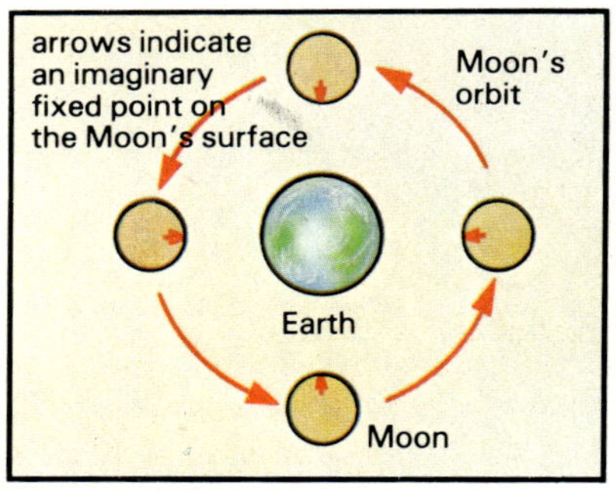

Copernicus is one of the younger craters and has a bright ray system.

The craters Alphonsus, Arzachel and Purbach are about 100 to 120km across.

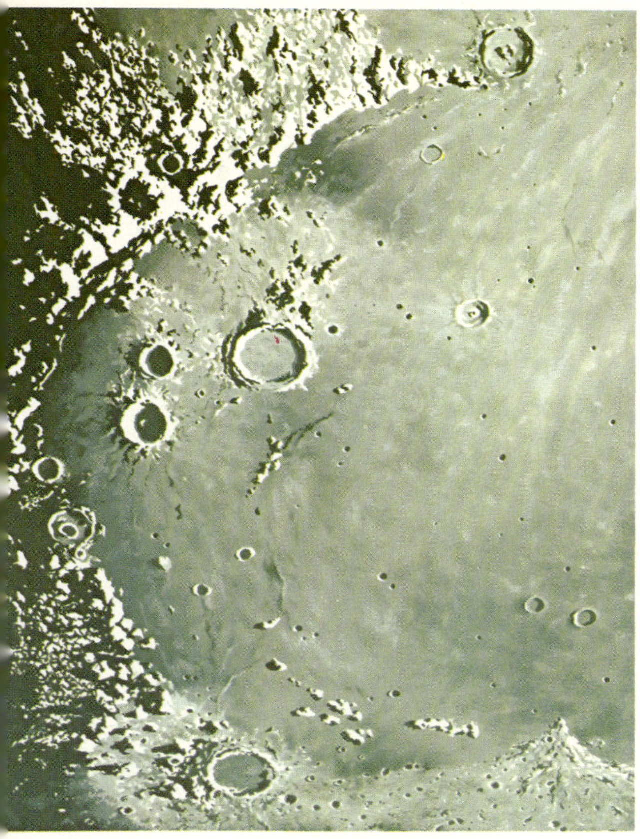

The Mare Imbrium is typical of the large, fairly flat, volcanic plains upon the Moon.

Below left: the far side, some of which is seen here, has few maria, but many craters.

Right: the Moon could have a small molten core *(top)*, but seismic results suggest it is mostly rigid with a weak inner region *(bottom)*.

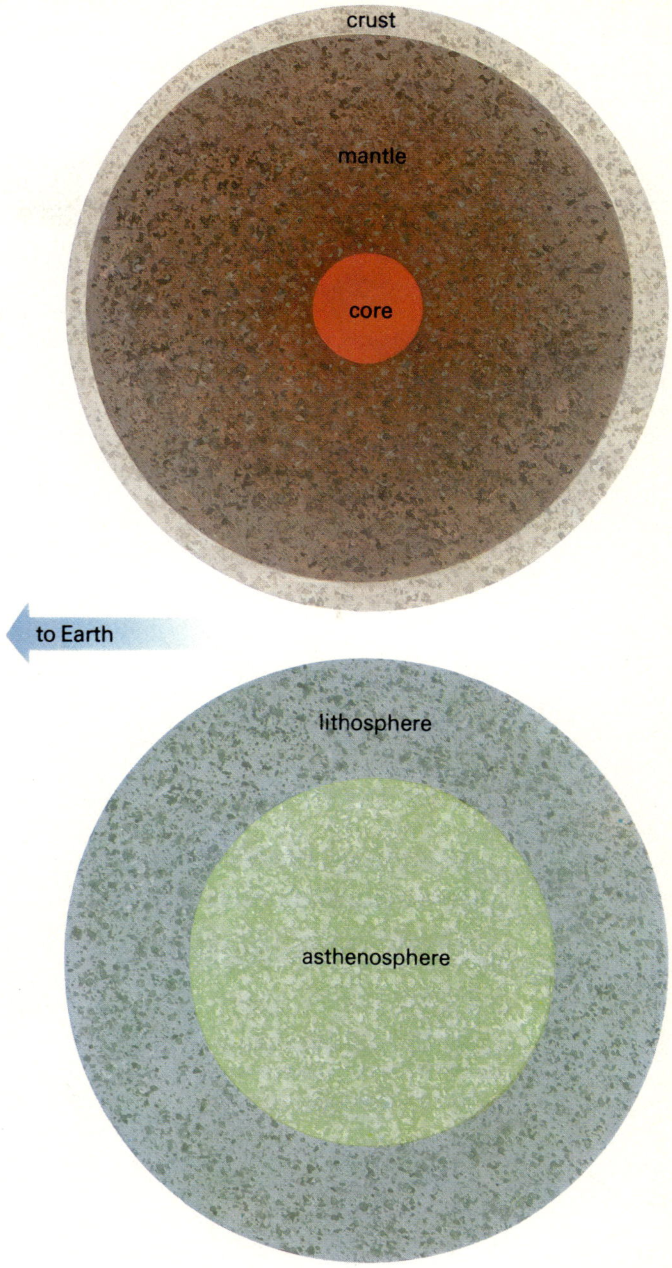

Below left: occasional obscurations of the lunar surface may be caused by gases escaping from the interior.

A view of Copernicus from an Apollo craft in orbit around the Moon.

Exploring the Moon

The results of our exploration of the Moon have been of the very greatest importance for our understanding not just of the Moon itself, but also of the whole history of the Solar System. Undoubtedly the most important single contribution has been the return of lunar samples by the Apollo programme, but many other pieces of information have been of great value.

Examination of the rock samples has shown that there are two basic types, one (known technically as an anorthosite)

The Apollo missions carried out many scientific experiments, like this one which collected particles from the solar wind.

Below: **the lunar rocks all proved to be very ancient, with none younger than about 3·1 thousand million years.**

The automatic Soviet Lunakhod vehicles moved around on the surface of the Moon and remained active for many months.

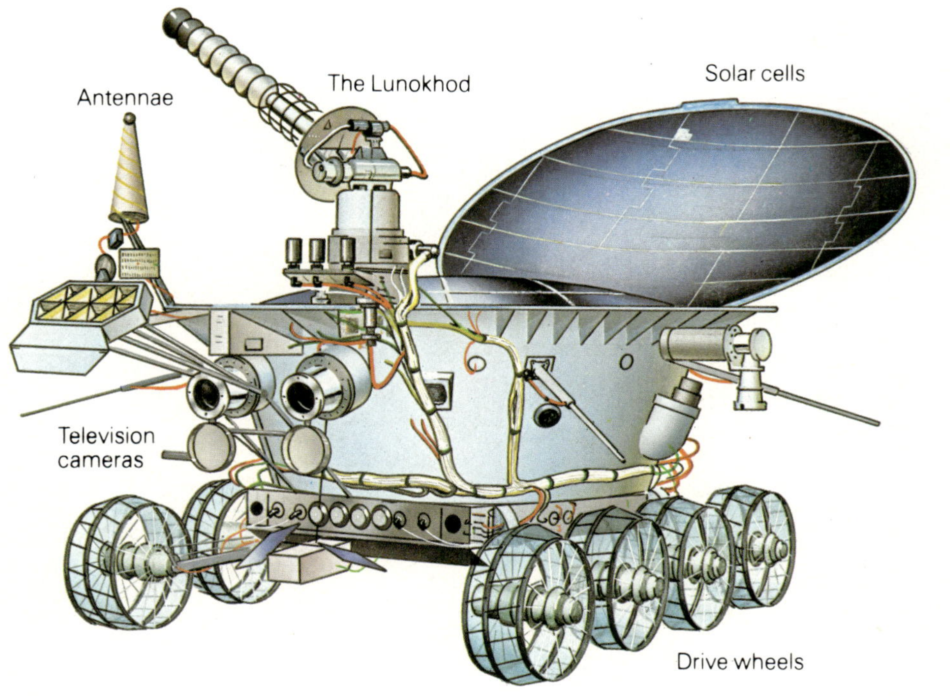

which apparently forms most of the lunar highlands, and basalt, which was once very fluid, covering the maria in a series of lava flows. Although the basalts are like similar rocks found on Earth, the overall composition of the Moon seems to be very different. This suggests that the Earth and the Moon were never part of the same body, but rather that they formed separately alongside one another, in the same region of the Solar System.

Seismometers installed on the surface showed that the Moon has a very thick rigid shell, and that only the very deep interior could still be molten. The very weak 'moonquakes' observed appear to be simply due to changing tidal forces as the Moon's distance from the Earth alters.

The sites of the Apollo landings upon the Moon.

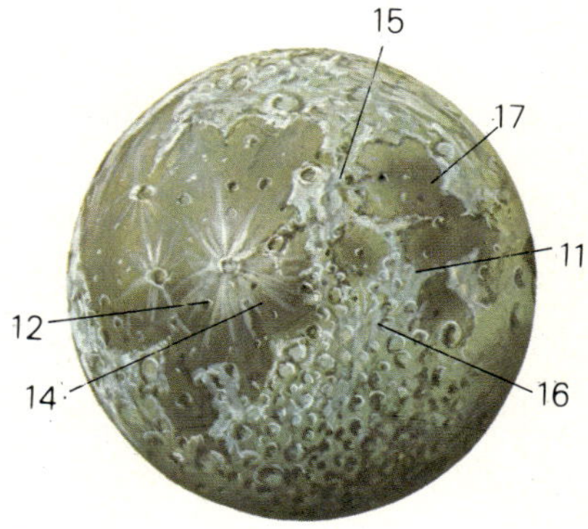

These large fractured rocks were examined in detail by the crew of the last Apollo mission– Apollo 17.

When the lunar surface was examined in detail it became obvious that the range of crater sizes was quite astounding, from the tiniest microscopic pits on lunar rocks to gigantic impact basins. Impacts had thrown up rocks from below the surface, welded some together and broken them to fragments again, and done this not once, but many times. The whole top layer had been reduced to a form of 'soil' – the regolith – consisting of rock fragments.

This alone should have been enough to convince anyone that the impact mechanism was important, but then, after studies of the rocks themselves, came the realisation that early in the Moon's history impact heating had produced a completely molten crust, and that this solidified to become the rocks of the lunar highlands. After this, further bombardment produced many of the craters seen today, while later still, volcanism flooded the areas now forming the maria. Radioactive dating techniques have enabled all this to be established, and have put the age of the oldest rock at rather more than $4\frac{1}{2}$ thousand million years – very close to the accepted age of the Solar System itself. On the other hand the youngest rocks are still about three thousand million years old – not much less than the oldest rocks known on Earth. Little has happened on the Moon since that time, whilst the Earth has changed beyond recognition.

Mars

To the naked eye, although Venus is brighter, Mars is probably the most striking of all the planets because of its distinctly red colour. Being the first planet beyond the orbit of the Earth, it can be opposite to the Sun in the sky, and at such times it is particularly well placed for observation. However, its orbit is fairly eccentric so that its distance from the Earth varies greatly, and some oppositions are far more favourable than others.

Telescopic observations reveal some de-

The earliest drawing of Mars, made in 1659 by Christiaan Huygens, compared with a photograph made through a large telescope.

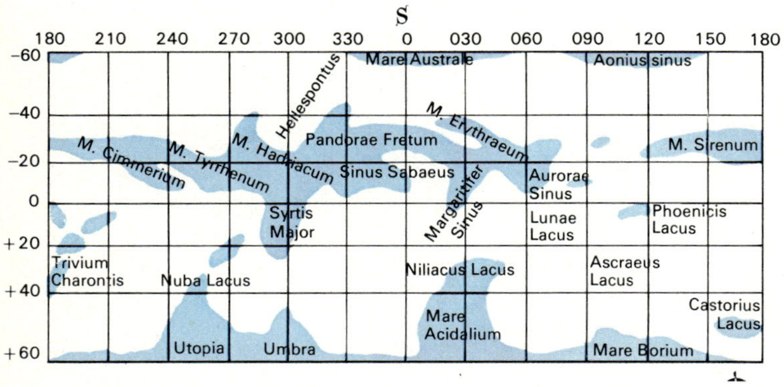

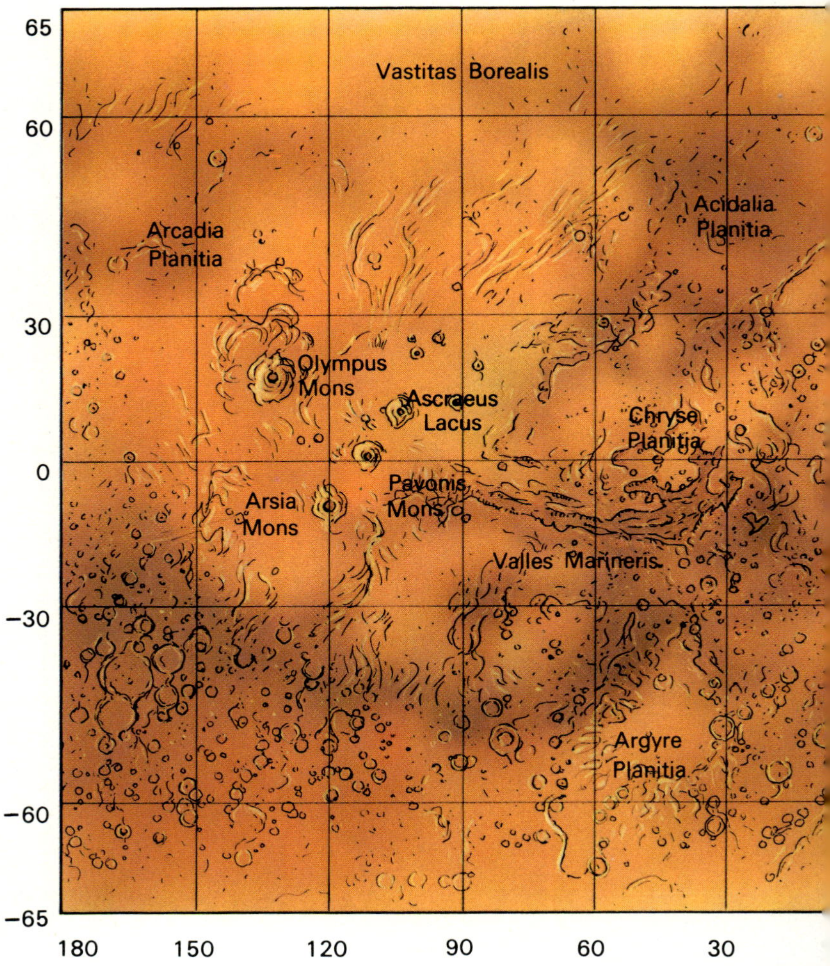

Above: **a typical map of Mars produced by amateur observers of the planet.**

Below left: **two of the many fanciful drawings which showed the purported 'canals'.**

Below: **the Viking spacecraft flew close to the moon Phobos, showing it to be a small, cratered lump of rock.**

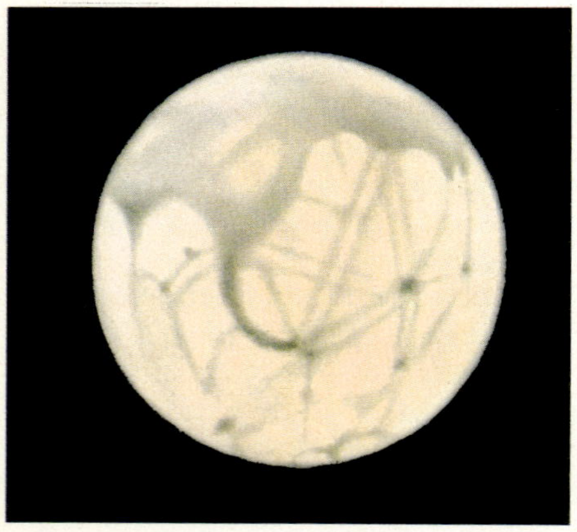

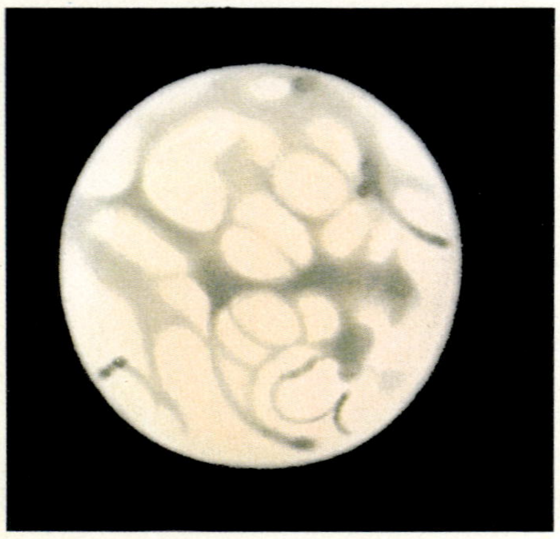

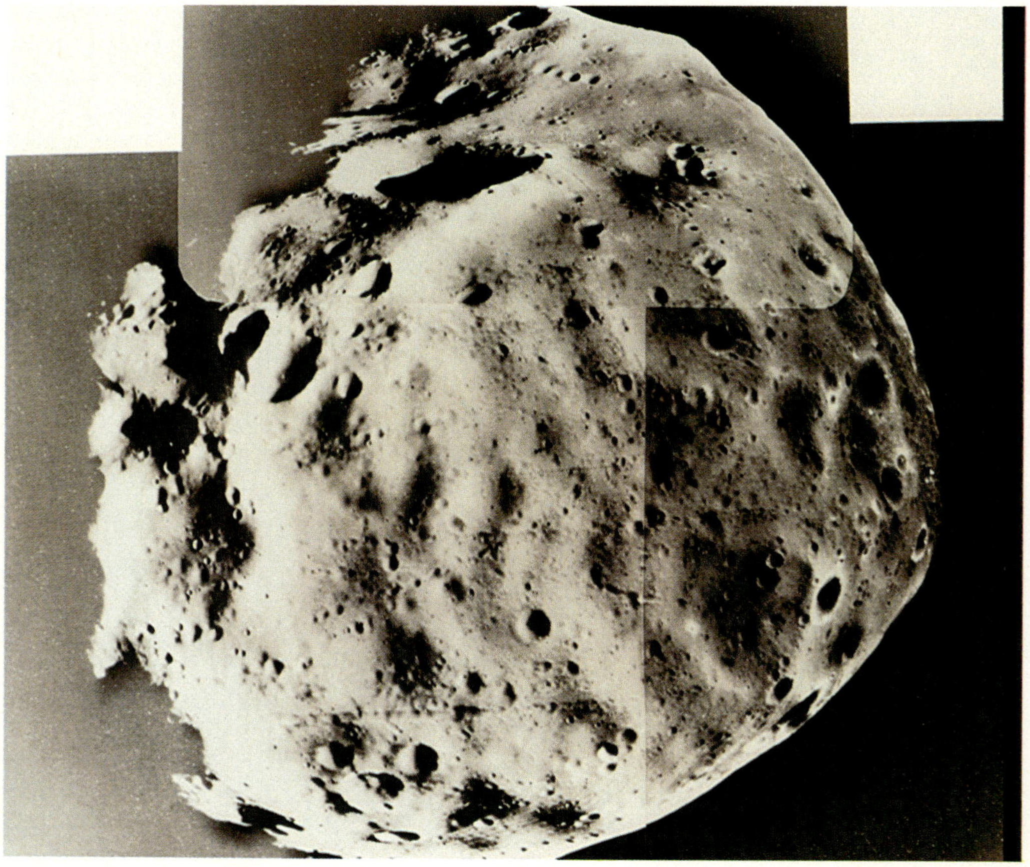

tails of the surface, including various dark markings, some of which are permanent while others slowly change over the years. For a long time there was argument over the existence of the so-called 'canals' – apparently straight lines criss-crossing the surface – but these are now known to have been completely illusory, the result of poor observing conditions and inadequate telescopes. The polar ice caps are usually visible, and these expand and contract with the seasons. 25km high is the largest such mountain anywhere in the Solar System. Just as striking is the vast Vallis Marineris rift valley system, which stretches for 4000km around the planet and is several kilometres deep.

The dust-storms had prepared astronomers for the signs of wind erosion which the spacecraft showed, but what was unexpected was the extensive evidence that Mars had once had running water, and that this had left dry channels and drainage patterns over wide areas of the surface. The existence of running water would be quite impossible under the thin atmosphere (mainly carbon dioxide) which exists today. The channels are very old however, and possibly at some time in the past the atmosphere was very much denser. There are signs that in some places a thick layer of ice may have existed (or may still exist) just beneath the surface.

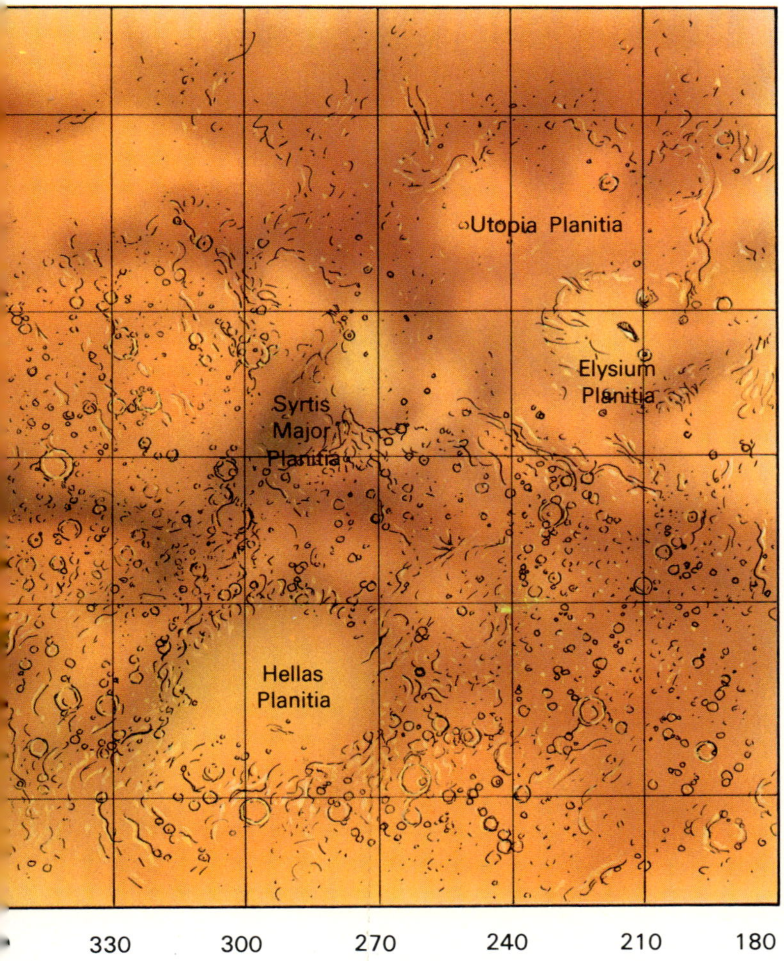

There are indications of thin clouds and occasionally dust storms, some of which may even extend planet-wide and hide the surface features for many months.

Mars is not very large – only 6795km in diameter – and it has an orbital period of 687 days and a 'day' which is very similar in length to that of the Earth, being just over 24½ hours. It has two tiny satellites, Phobos and Deimos, orbiting close to the planet, and too faint to be seen by any but the very largest telescopes.

This was more or less all the information we had about Mars until it was visited by spacecraft. Then the situation changed dramatically. The first surprise came in 1964 when Mariner 4 found craters on the surface. Strangely enough, no-one had really expected these, although with hindsight they should have done. The next important mission, Mariner 9, mapped the surface and this work was continued by the two Viking Orbiters. An immense wealth of detail was revealed, very little of which bore any relationship to the features charted from Earth. As it happened, the craters were probably the least interesting of all these new details. There are several large volcanoes, for example, and one, Olympus Mons, at over

Above left: **only in a few cases do the markings visible from Earth bear much relationship to the features mapped by spacecraft.**

Right: **numerous surface features, as well as atmospheric hazes are seen in this Viking photograph of Mars.**

Below: **the two tiny satellites of Mars orbit close to the planet.**

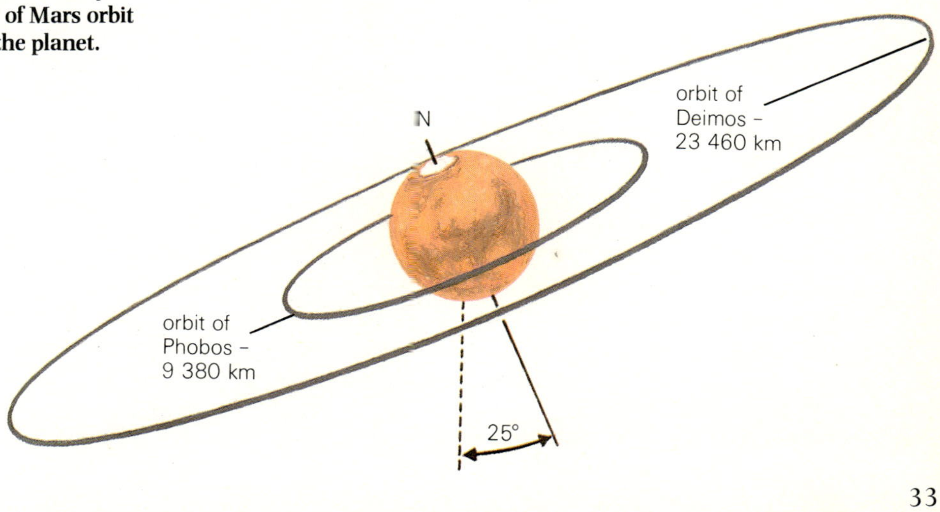

The surface of Mars

Quite apart from the suggestions that the (non-existent) 'canals' might have been built by some form of intelligent beings, there had been other possible indications that Mars might support some form of life. As the polar caps melted after the winter there seemed to be a 'wave of darkening' which spread outwards towards the equator, and it was thought that this might be due to the sudden growth of vegetation with the coming of spring and a supply of available water. It is now recognised that this was partly an illusion, and partly caused by an alteration in the pattern of the Martian winds which sweep certain areas clear of surface dust. The changes in the features noted from Earth are now known to be caused by similar shifts in the distribution of the fine wind-blown particles which cover many areas of the globe. It is the winds, too, which have been responsible for shaping the many sand dunes observed on various parts of Mars, including the vast region, larger than all the deserts of the Earth, around the northern polar cap.

The ice-caps have proved to be of quite ordinary water ice, although in the depths of winter the polar temperature becomes so low that some of the carbon dioxide from the atmosphere freezes out on to the surface, and in such quantity that it actually alters the atmospheric pressure. Detailed photographs show that the northern cap has a strange laminated structure. It is probable that this is due to alternating layers of dust and ice, and it may well preserve a record of various climatic changes.

Most of the clouds and mists consist of water droplets and of ice, but some of the very high-altitude clouds are carbon dioxide ice. Just as on Earth, the mists form overnight in low-lying valleys and disperse with the warmth of the morning sunlight. Extensive cloud-sheets may sometimes build up over the high mountains, and spiralling cloud patterns show that Mars too has its depressions. But the atmosphere and the surface are generally very dry, and although there may be frosts, no rain can ever fall.

Most of our information about surface conditions has come from the two Viking Landers which reached the planet in 1976. These returned photographs showing that the landscape at one site was gently undulating, and at the other a fairly flat plain, but that both were covered by rocks and fine-grained soil. The rocks appear to be volcanic, although rather different in composition from terrestrial rocks, and iron compounds give them their rusty colour.

Needless to say, no vegetation or higher life forms were discovered, but the Viking Landers also carried equipment which could detect the effects of microscopic life-forms, and make tests for organic compounds. At first there was excitement when one experiment apparently indicated some form of biological activity, but it is now agreed that rather unusual (and purely chemical) reactions were the cause.

Right: a model of the enormous volcano known as Olympus Mons.

The Viking Lander vehicles showed the Martian surface to be undulating plains covered in reddish rocks.

Sunset on Mars with _(inset)_ early-morning mists in the valleys of Noctis Labyrinthus.

Right: the Viking Landers were highly complex instrumented packages only about 2m across.

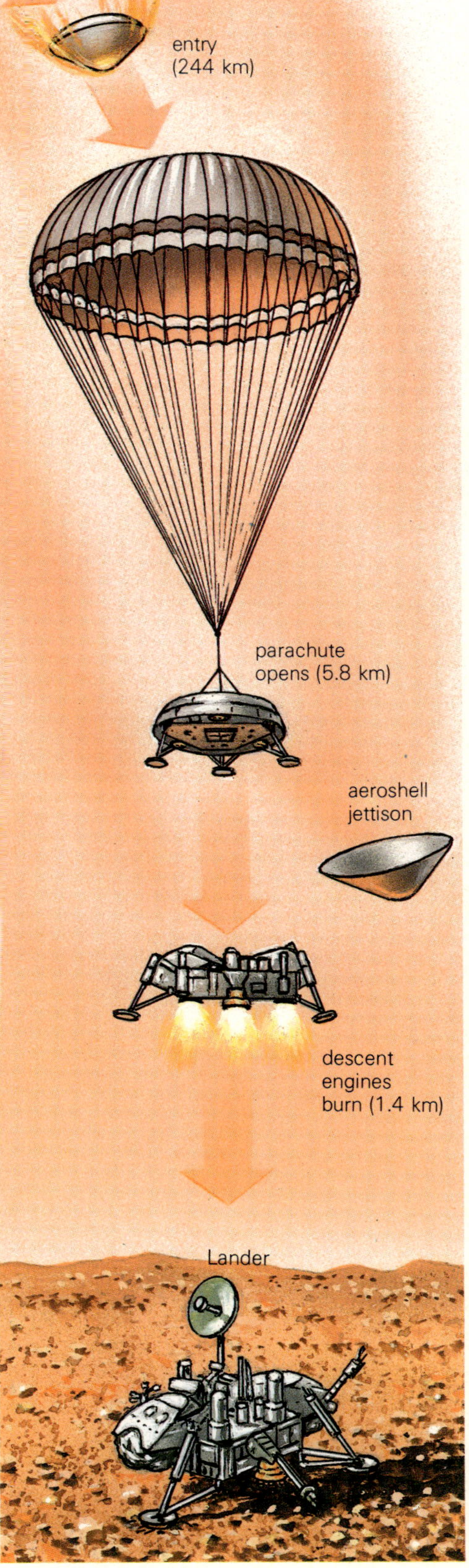

The Viking Landers parachuted down onto the surface, making atmospheric measurements on the way.

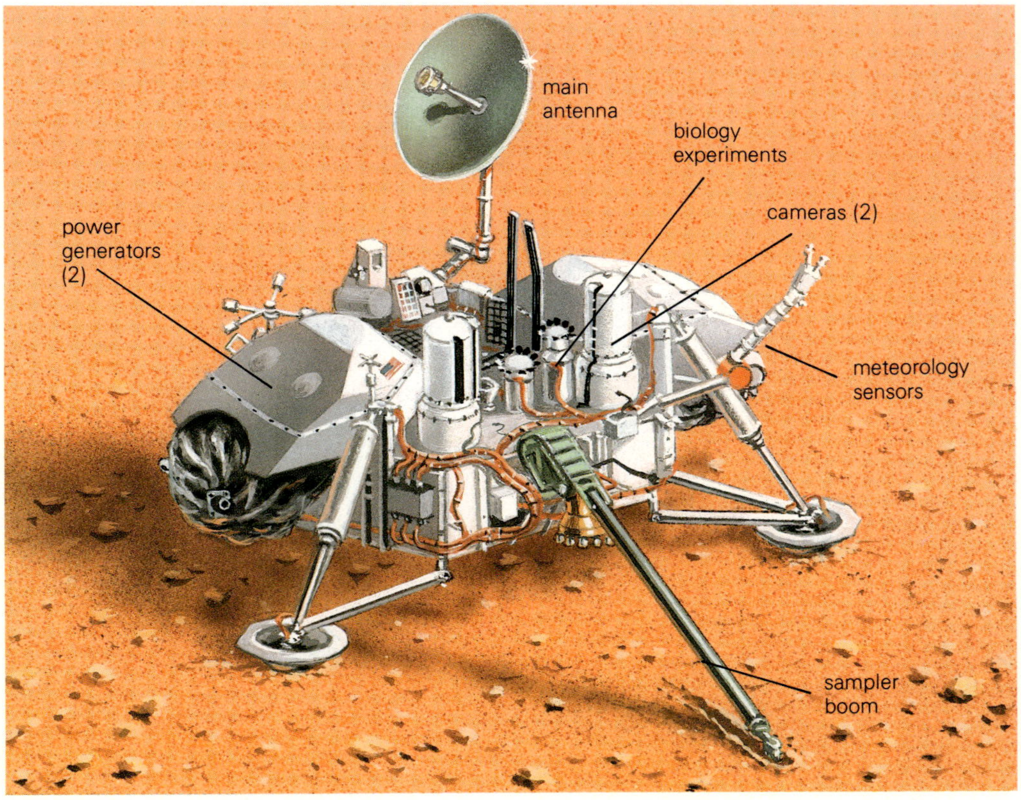

Minor planets and meteorites

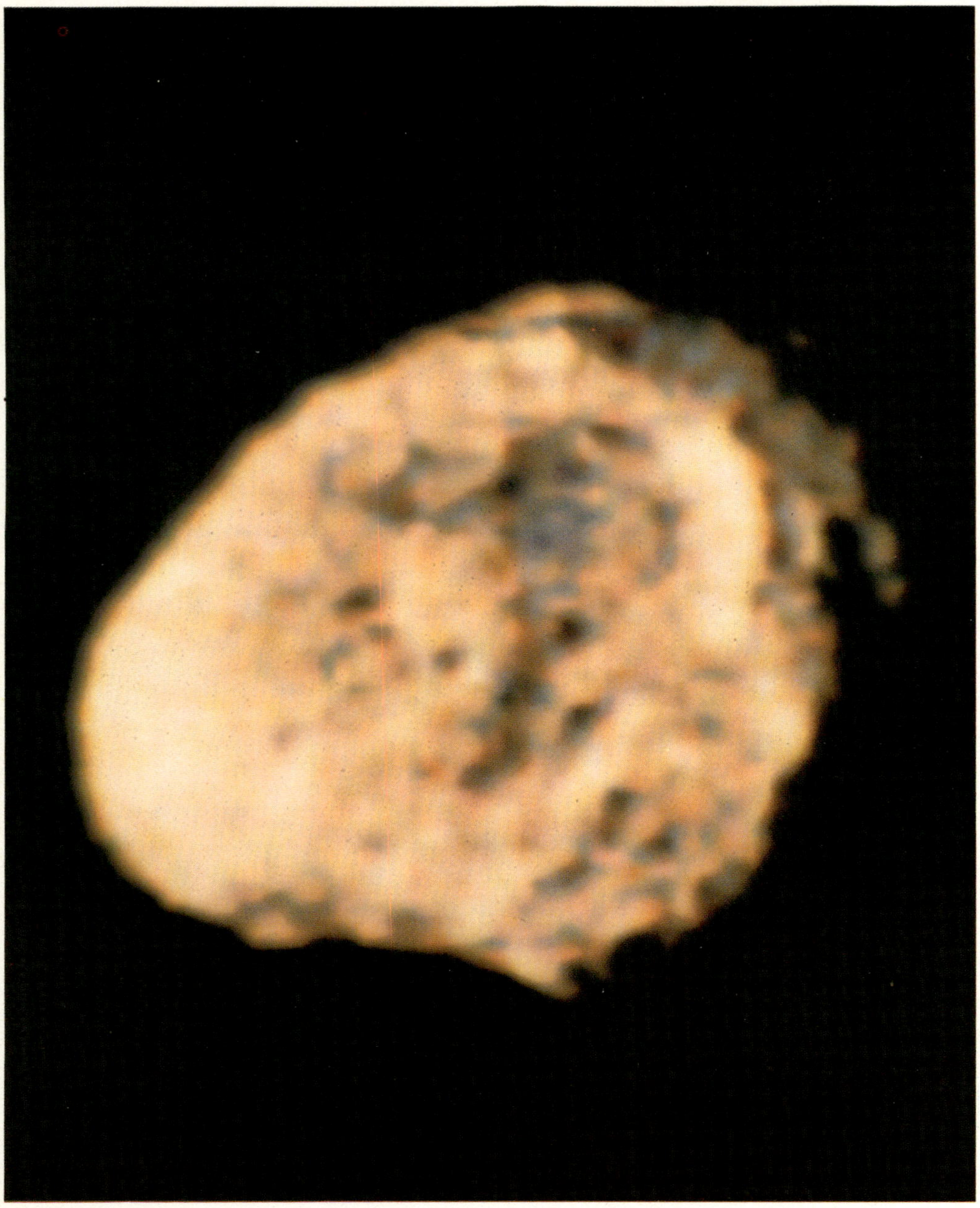

Apart from the nine major planets the Solar System contains large numbers of smaller bodies, most of which have orbits between those of Mars and Jupiter. Roughly three thousand **minor planets** are known, but there may be hundreds of thousands altogether. Most are very small, little more than lumps of rock; even the largest, Ceres, is no more than about 1000km across, and only one, Vesta, can ever be seen by the naked eye.

Quite a number of minor planets are known to come inside the orbit of the Earth, and some go closer to the Sun than Mercury. There must be many thousands more which are undetected. Of those that pass the Earth some stand a chance of hitting the atmosphere and of burning up in the form of a brilliant **fireball.** (The generally fainter meteors come from comets.) Part of the body may even survive to fall to Earth as a **meteorite.** Although most must be lost in the ocean, many meteorites have been recovered. Especially in recent years, many hundreds have been found in Antarctica, perfectly preserved by the cold conditions. (One of these has actually come from the Moon, as the rock is identical to that of the lunar highlands.)

Meteorites are of two main types, the stony and the metallic – the latter containing mostly iron and nickel. A much rarer type (known as the carbonaceous chondrites) includes a number of carbon compounds, and at one time it was suggested that these might have originated from some form of life. However, it is now known that this is not the case, and that the compounds are similar to the molecules found in interstellar space, from which they probably derive. This type of meteorite appears to be much older and less altered than the others, and preserves a record of events which took place very early in the history of the Solar System, or even before it was formed.

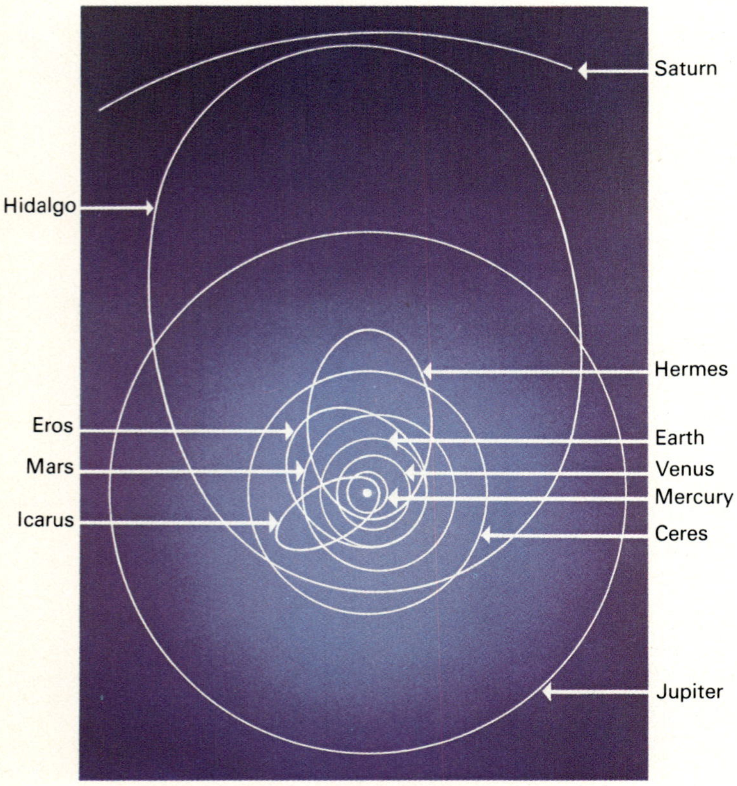

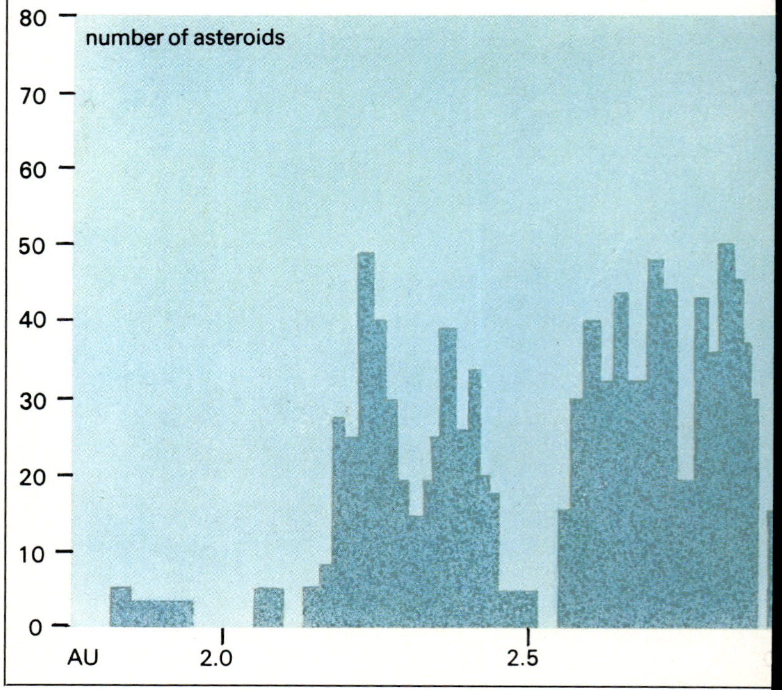

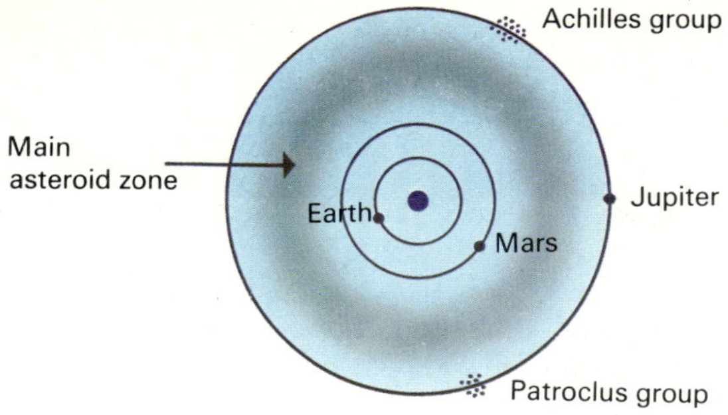

Soon after the minor planets were discovered it was suggested that they might be fragments of a single large body. Studies of meteorites and of the minor planets show that this cannot be the case, but that quite a number of fair-sized bodies were probably formed at the same time as the larger planets. When they were fragmented by collisions they gave rise to the minor planets and meteorites which exist today. The stony and metallic bodies are found closest to the Sun, but further out many minor planets resemble the rare carbonaceous chondrites, and others are like the icy moons of the major planets. When the Solar System was formed it must have included many icy bodies, but only those far from the heat of the Sun survived.

Opposite top: **Hyperion, a satellite of Saturn, is probably very similar to many of the minor planets.**

Opposite bottom: **the orbits of minor planets show much greater variation than those of the larger bodies in the Solar System.**

Above: **most minor planets orbit between Mars and Jupiter, with some in the same orbit as the giant planet.**

Below: **the distribution of minor planets shows many 'groups' and 'gaps'.**

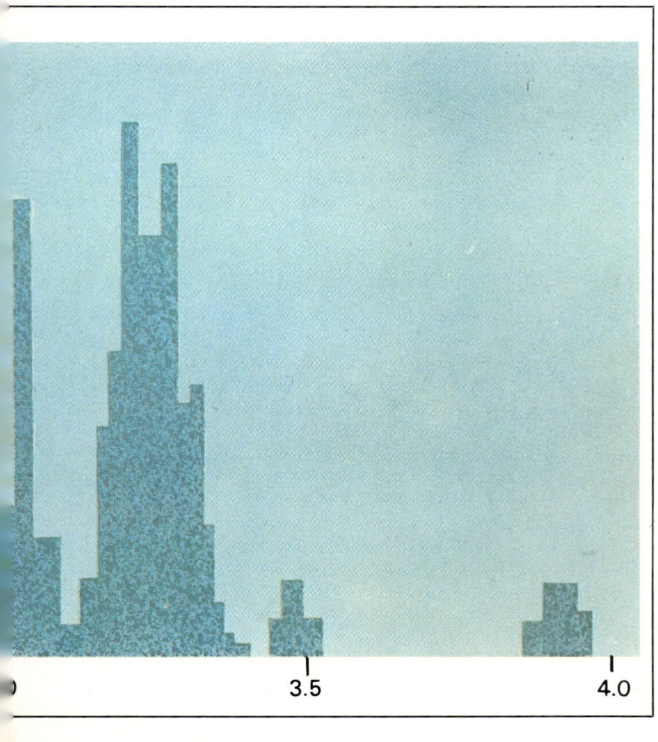

Stony meteorites are most numerous, but less frequently recovered than the irons.

One of the Greenland meteorites *(left)* **weighs about 36 tonnes.** *(Right)* **A cut and polished section of an iron meteorite.**

Impact craters

Most meteorites reaching the surface of the Earth are quite small and have been braked by atmospheric friction, losing most of their speed at some height above the ground. They then fall more or less vertically and dig a comparatively small hole in the surface. This can happen with quite large bodies: the heaviest known intact meteorite, an iron, found near Grootfontein in Namibia, is thought to weigh more than 60 tonnes.

Any object approaching the Earth must have a velocity of at least 12km per second – it is usually much more – and if it is large enough to hit the ground at about that speed, the result is very different. The energy is released in a true explosion, which vaporizes the meteorite and part of the surface. Frequently, layers of rock are turned back over the surrounding surface, so that they come to rest upside down. Other rocks may be flung out for considerable distances, and on an airless world such as the Moon, small particles may travel for hundreds of kilometres and produce bright ray systems around the crater. Only very rarely can any material from the meteorite be discovered.

Most of the planets and their satellites show signs of this sort of bombardment. Even the Earth, which is very active geologically, and rapidly erodes such craters, bears quite a number of scars left by ancient impacts. Some of the satellites of Jupiter and Saturn appear to have had some impact craters removed by geological processes, and Venus has probably escaped the worst effects because of its dense atmosphere which would burn up quite large bodies. But Mercury, and many satellites – the Moon, Callisto and Rhea are examples – are covered

Artist's impression of a crater-forming impact which vapourizes part of the original surface.

Left: **the surface of the Moon is densely covered with impact craters.**

Opposite top left: **meteorites undoubtedly come from space, but tektites are probably formed from terrestrial rocks.**

Opposite top right: **tektites occur only in specific areas of the world.**

Opposite centre: **Meteor Crater in Arizona is about 1km across, and was formed about 20 000 years ago.**

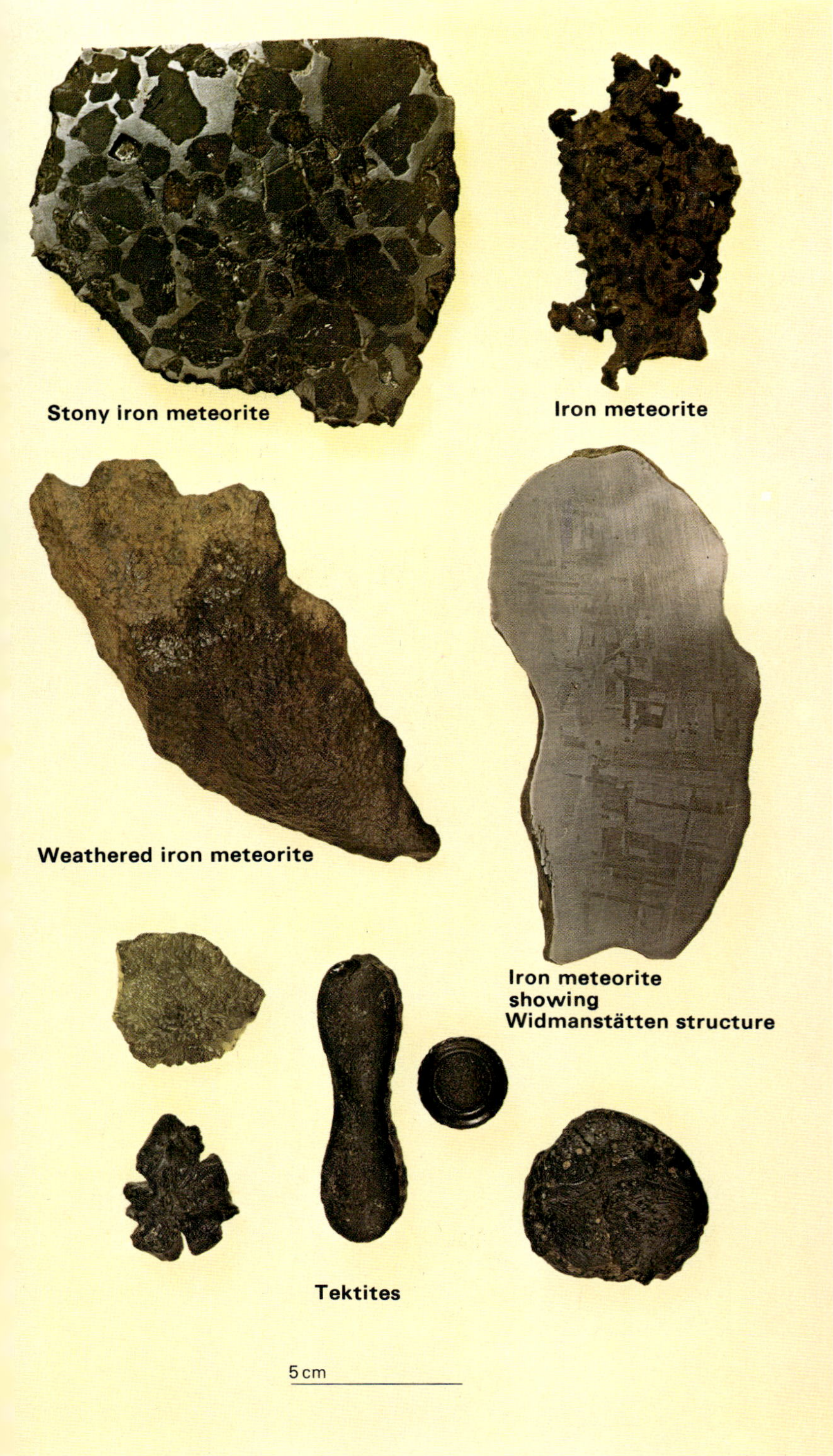

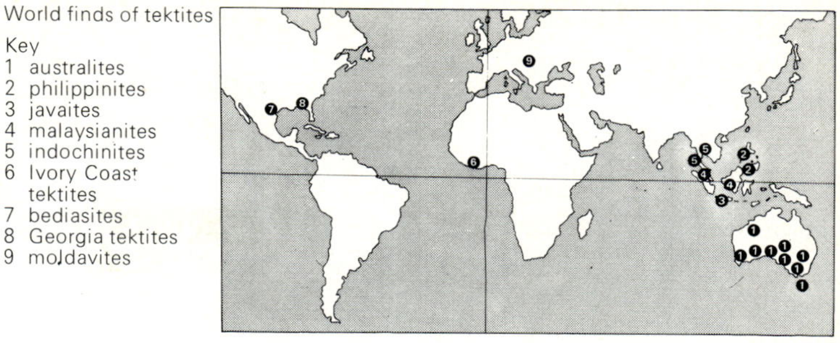

in craters, sometimes so thickly that any new craters would destroy as many old ones. Most of these impacts are very ancient, probably produced before the Solar System was a thousand million years old. Although impacts still occur, they have become very infrequent. The famous Meteor Crater in Arizona is about 20 000 years old, and the last true explosion crater of Earth was probably formed in Estonia about 4000 years ago. (The enormous explosion which devastated part of a remote Siberian forest in 1908 is thought to have been due to part of a comet.)

Meteoritic impacts may explain the mysterious glassy objects called **tektites** which are found in several regions of the world. They have clearly been molten and shaped by their passage through the air, but no-one knows their origin. It has been suggested that they come from the Moon, but they are unlike lunar rocks and it is more likely that they were formed by unknown terrestrial impacts.

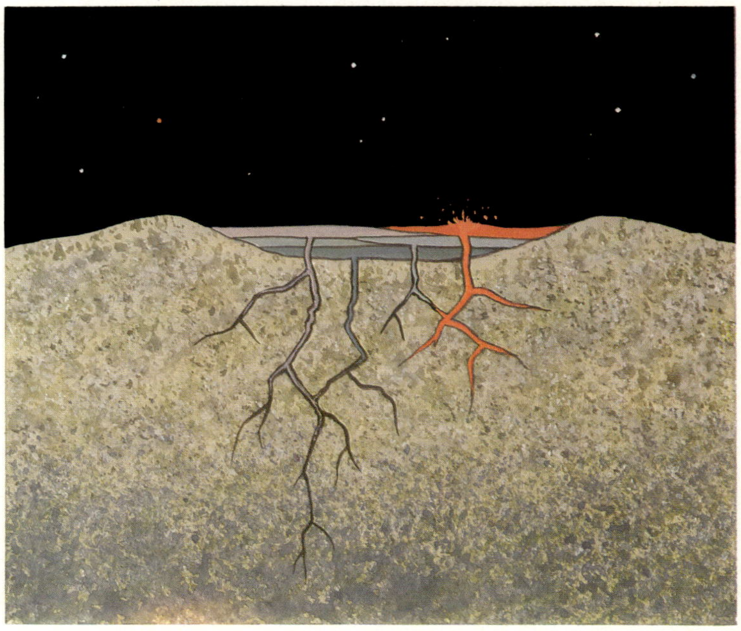

Many lunar maria were formed as impact craters, the floors of which were flooded by later lava flows.

Jupiter

The largest planet in the Solar System, with a diameter of 142 800km and as massive as 1340 Earths, Jupiter consists mostly of hydrogen and helium, but probably has some rocky material at its centre. Most of the interior is liquid and there is no definite surface below the dense atmosphere of ammonia and methane, where only the topmost layer of frozen ammonia crystals can be seen. Other layers further down consist of ammonia compounds, and, deeper yet, ice and water droplets.

Broad bright zones of rising air and dark descending belts encircle the globe, while strong winds of more than 500km per hour blow round this fast-rotating world (which spins once every 9 hours 55 minutes). The tremendous weather systems include all sorts of small spots, streaks and whirls, but only a few large white ovals and the Great Red Spot are very long-lasting. The Great Red Spot is huge – 25 000km long and larger than the whole surface of the Earth – and has been seen for over 300 years, although it is gradually shrinking and changing colour.

This whirling, restless atmosphere is lit up at night by powerful lightning strokes, producing radio waves which can be detected even as far away as Earth. One of the Voyager spacecraft also found vast auroral displays caused by the planet's huge and powerful magnetic fields.

Around the planet is a thin ring of tiny particles, its outer edge at 124 400km from the centre, and probably reaching down to the tops of the clouds themselves.

Jupiter has at least seventeen satellites, most of which are small and unimportant, although four (or more) of the outer ones may be captured minor planets. Yet four moons are the size of planets. Callisto, the furthest out, has a cratered surface and the remains of one enormous impact which sent out ripples for more than 1 500km around it.

Ganymede is the next in, and like Callisto it

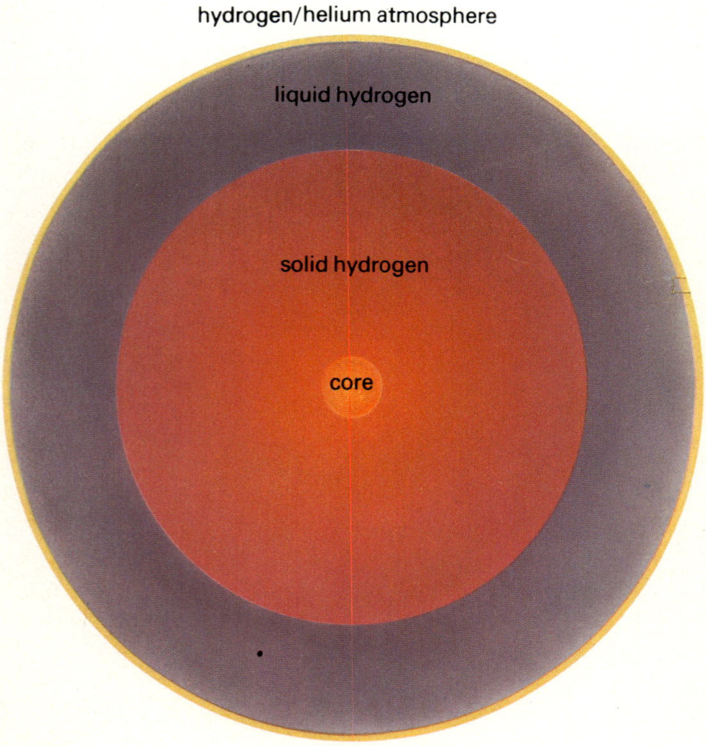

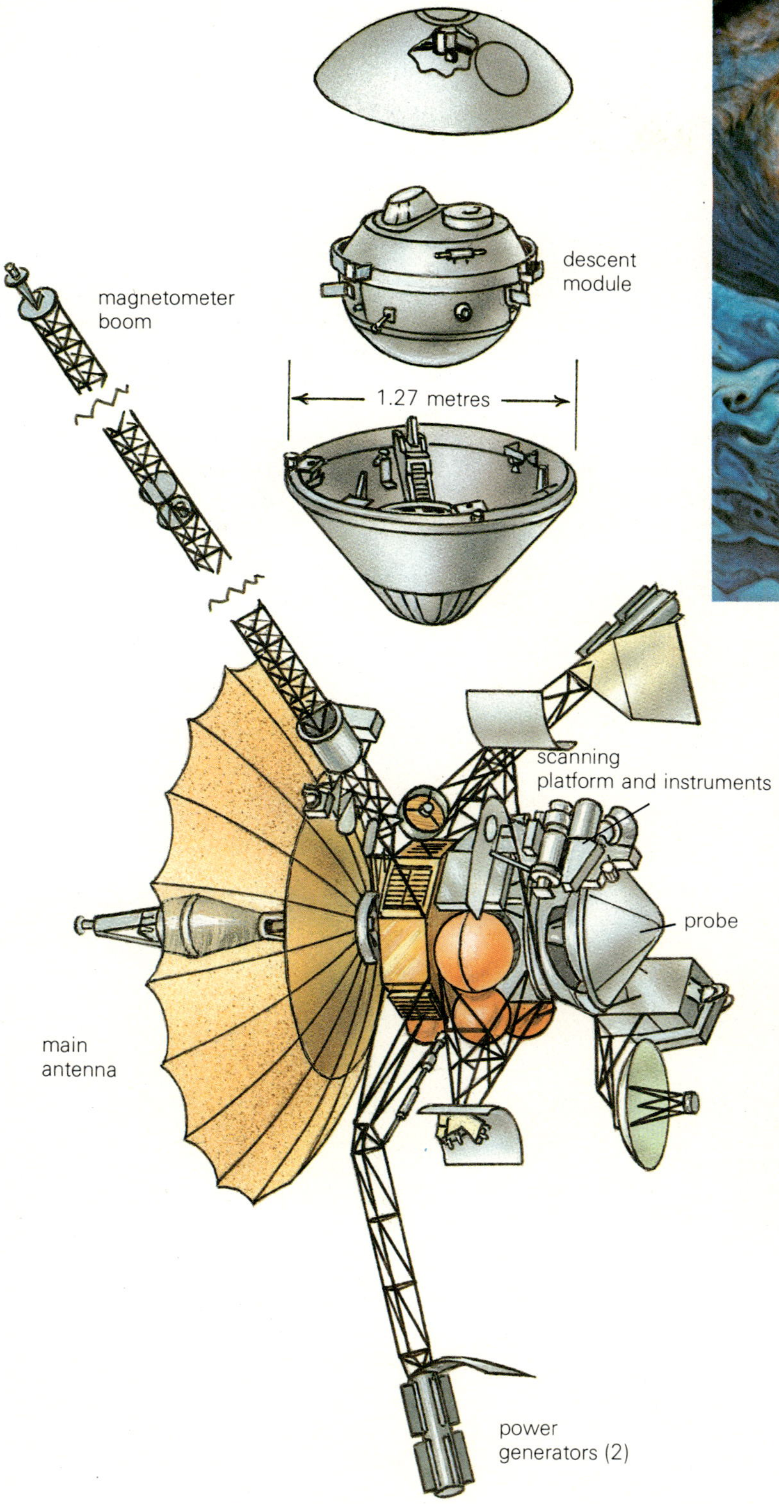

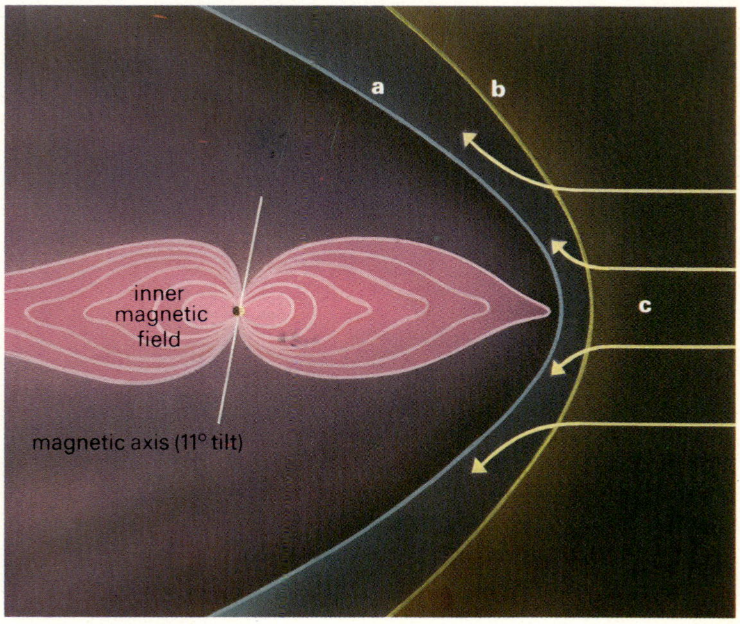

Top: **this 'false-colour' photograph of the Great Red Spot shows up some of the features more prominently.**

Above: **a bow shock wave (b) is produced where Jupiter's magnetosphere (a) encounters the solar wind (c).**

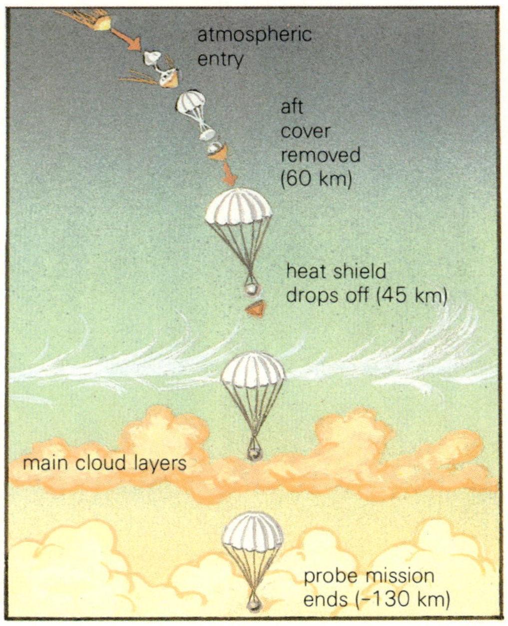

Opposite top: **this photographic composite shows Jupiter and its major moons, Io, Europa, Ganymede and Callisto.**

Far left: **in the interior of Jupiter, the solid hydrogen acts like a metal.**

Left: **the turbulent atmosphere of Jupiter, showing the Great Red Spot, and some of the white ovals.**

Above and right: **the Galileo spacecraft is intended to study Jupiter and the satellites for an extended period.**

probably has an inner core of rocky material surrounded by shells of liquid water and ice. The largest satellite in the Solar System, its diameter of 5276km is more than that of Mercury. Strange 'grooved terrain' looking like a harrowed field separates older, darker plains.

Even further in, Europa is very strange – like a smooth round ball covered with dark marks and low white ridges. The surface must be very new because few craters can be seen, and everything has probably been evened out by 'volcanic' cracks throwing water out on to the icy surface.

The oddest world of all is Io, which is red, orange and white in colour. Volcanoes, more active than those on Earth, erupt not lava, nor even water, but liquid sulphur and its compounds. A small globe like Io should be cold, but the giant planet Jupiter squeezes it with tides and heats the interior, and it is this that gives rise to the volcanoes.

Saturn

Saturn is another giant planet, 120 000km across and flattened by its fast rotation. Its day is only 10 hours and 40 minutes long, but it takes nearly 29½ years to complete one orbit around the Sun. In the thick atmosphere a few belts and spots can be faintly seen from Earth, but a heavy haze hides most of the clouds of ammonia crystals. From close at hand the Voyager space probes detected many features, and although the weather is generally less violent than Jupiter's, they did discover that at the equator wind speeds can reach 1800km per hour.

Other planets have rings, but none can compare with those of Saturn for size and beauty. From Earth only three can be seen, together measuring about 70 000km wide but only a few kilometres thick. But how many rings are there? The Voyagers found that the main rings are really thousands of tiny ringlets and that not even the divisions and gaps are truly empty – they too have tiny bodies and faint ringlets within them. There must be thousands of millions of particles in the rings, all of them consisting mostly of plain water ice, and ranging from tiny fragments to blocks 10m or so in size. No-one is sure whether they are material left over from the formation of the planet, or are fragments of a larger satellite.

Saturn has at least as many satellites as Jupiter – seventeen – and may have several more. One, Titan, is large (5150km in diameter), and is the only known satellite with an atmosphere. Surprisingly, this is mostly nitrogen and the pressure is greater than on the Earth. The surface is hidden, about 250km below a thick hydrocarbon haze which resembles a heavy pollution smog. Titan is cold (–182°C), and methane rain falls on a dark and icy surface.

Six of the other satellites are larger than 400km across and all are mainly ice, with quite heavily cratered surfaces. One, Iapetus,

Left: **this montage of photographs shows Saturn with** *(clockwise from top right)* **Mimas, Tethys, Dione** *(in foreground)* **and Rhea.**

Opposite below: **the Voyagers showed Saturn's rings to be composed of thousands of individual tiny ringlets.**

Below: **the main rings of Saturn with some of the inner satellites.**

Right: **before the visits by Pioneer and Voyager spacecraft this was one of the very best photographs of Saturn.**

Bottom right: **as seen from Earth the appearance of the rings changes greatly due to the planet's inclination.**

has always been a mystery because one side is white like ice, while the other half is nearly black. Did the dark coating come from outside, or from the moon itself? Even the Voyager pictures did not provide the answer.

Gravitational forces have locked several tiny satellites into the same orbits as those of large moons, but two perform the strangest dance around the planet. Roughly 200 and 120km across, these have nearly the same orbit, but one moves slightly faster than the other and gradually catches up with it. Before they collide, gravity swings them round one another, changing their places. The slower one becomes the faster, and begins a four-year chase around the orbit.

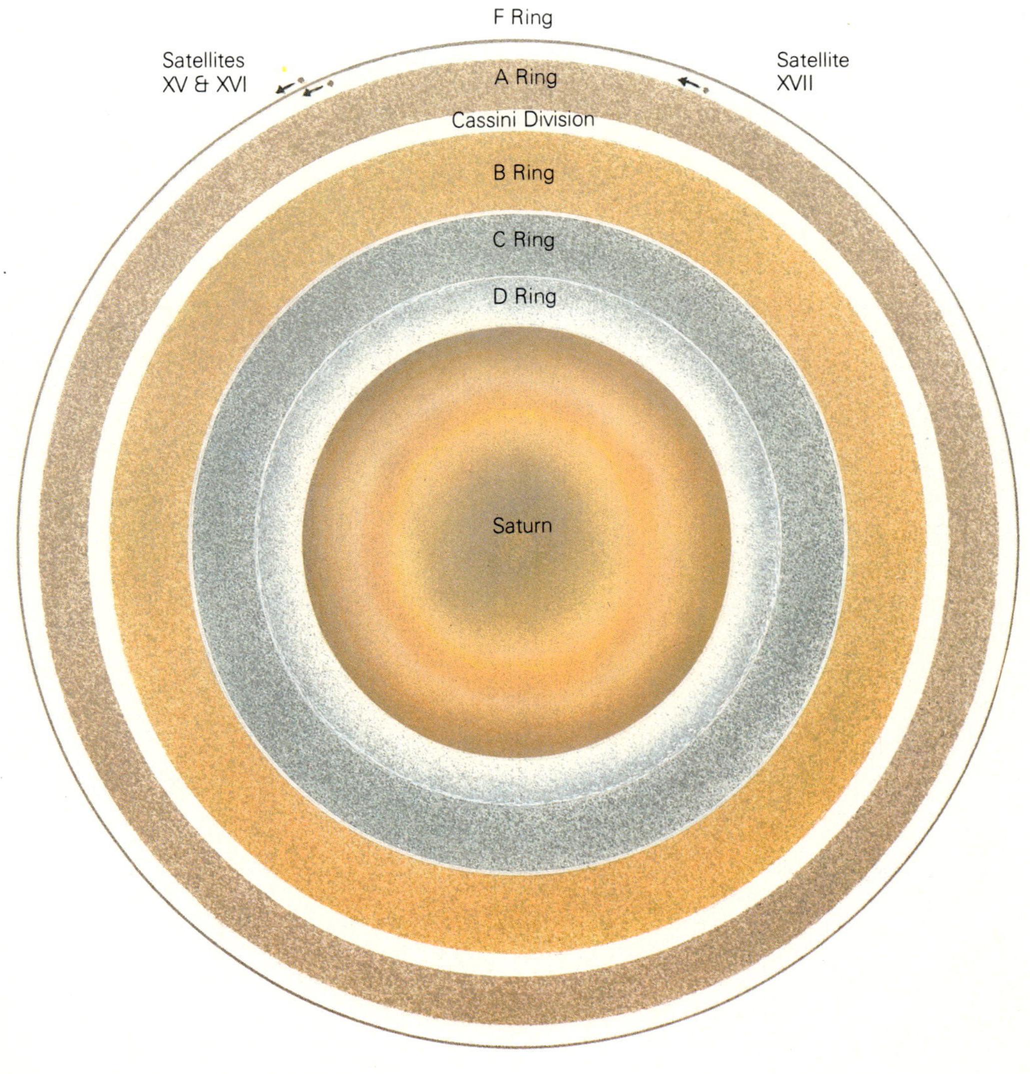

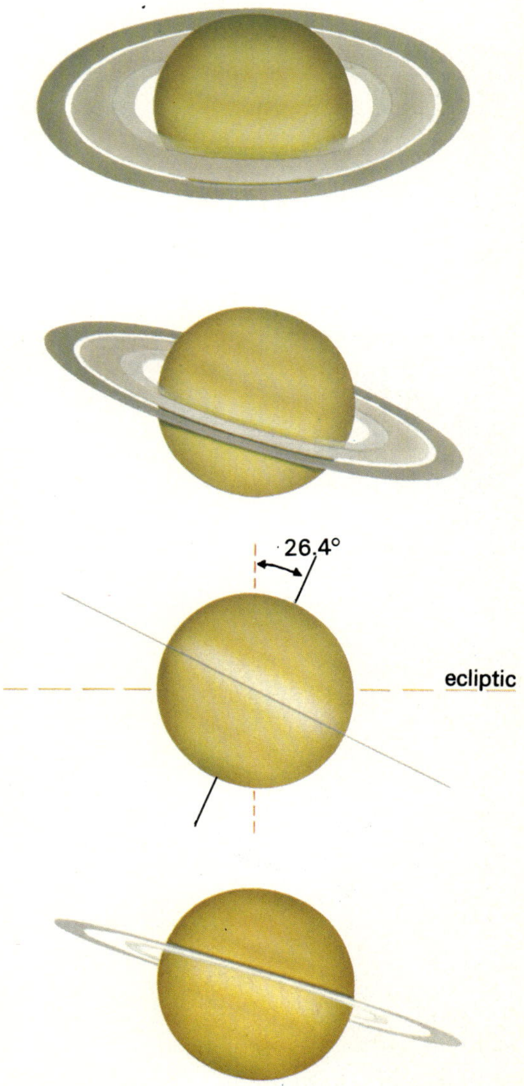

The outer planets

The two main planets in the outer Solar System, both 'gas giants' are like a pair of twins. **Uranus** is closer to the Sun and slightly larger, with a diameter of 50 800km to **Neptune**'s 48 600km. Both are close to 16 times the mass of the Earth, Neptune slightly more, Uranus slightly less. Uranus takes about 84 years and Neptune 164 years for one complete orbit, but the oddity of Uranus is that its rotational axis is tipped over by more than a right angle, so that the poles sometimes face the Sun.

Both planets have atmospheres of hydrogen and ammonia, with some methane probably included. Changes have been observed on Neptune which may be due to variations in clouds of frozen methane, but few details can be seen on either planet, although Voyager 2 may send back pictures of Uranus in 1986.

Five satellites and at least nine narrow rings of some very dark material orbit Uranus in the plane of its equator. Neptune might have a ring – observations are inconclusive – but definitely has two satellites, Triton the larger, 3700km across, and tiny 300-km Nereid. A possible third moon has yet to be confirmed.

The ninth planet, **Pluto**, has an eccentric orbit which sometimes (as at present) brings it closer to the Sun than Neptune, but most of its 248-year orbit lies well outside that of Neptune. Pluto might just have a methane atmosphere when closest to the Sun, but most of the time the temperature will be below –115°C, and it will be frozen on the surface.

After Uranus was discovered by William Herschel in 1781, it was found that some-

Right: **Uranus shows few surface features, but has several narrow, dark rings.**

An artist's impression of Neptune and its large moon, Triton.

Charon, Pluto's only satellite is nearly half the size of the planet itself.

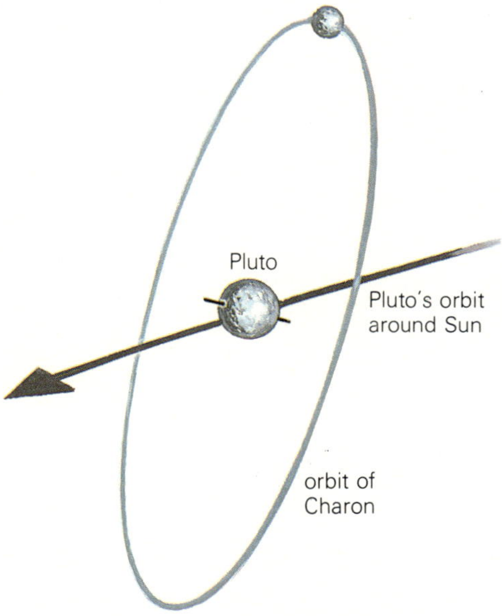

Right: the orbit of Charon is highly inclined to Pluto's orbit around the Sun.

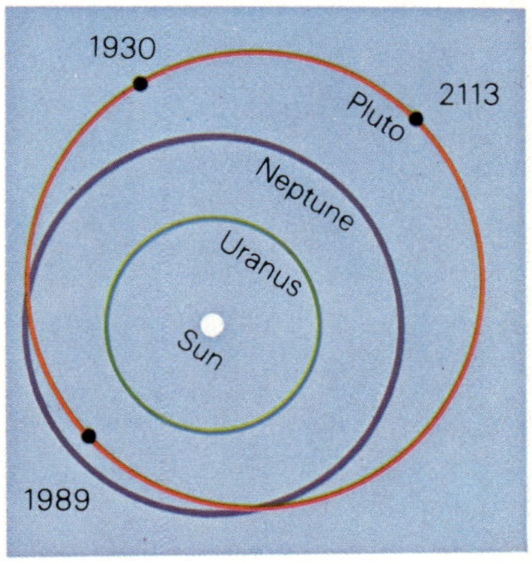

The orbit of Pluto is very eccentric, bringing the planet in within the orbit of Neptune.

The Pioneer 10 spacecraft is now further from the Sun than any of the known planets.

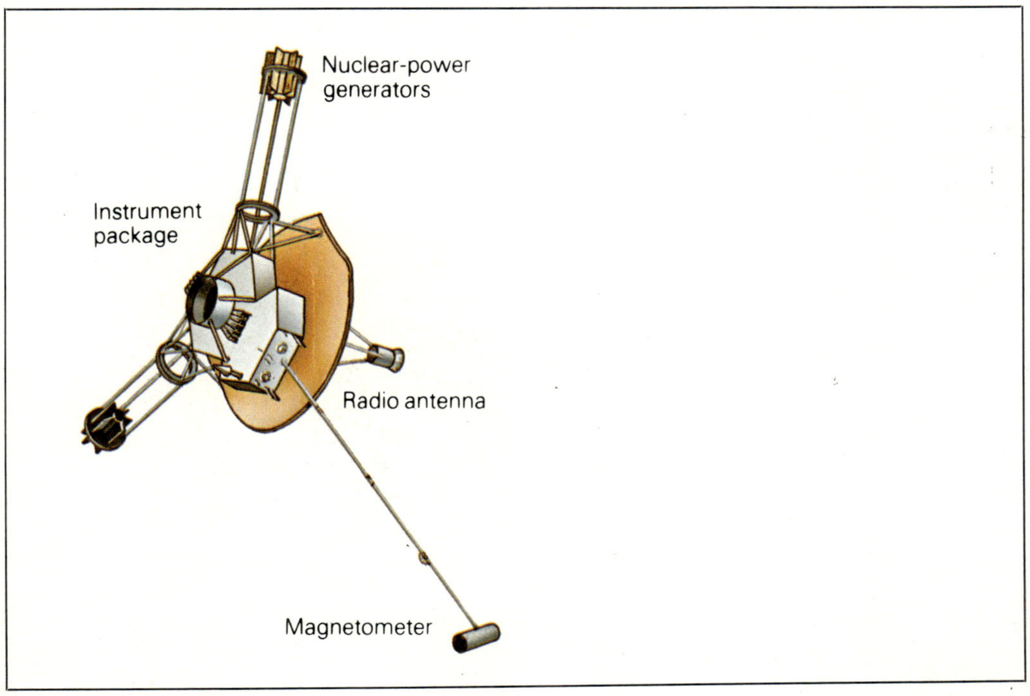

thing was pulling on the planet. Le Verrier in France (and Adams in Britain) made calculations which predicted another more distant planet. In 1846 Galle in Berlin found Neptune, but then that planet proved to be smaller and lighter (and closer) than expected, and the problem still remained. Later Percival Lowell in the U.S.A. started a photographic search, from which Clyde Tombaugh discovered Pluto in 1930. But it too is smaller and lighter than predicted, so the question remains – is there yet another planet out there? Perhaps it is not a planet, but a dark dwarf star, companion to our Sun, more likely than the black hole which has also been suggested. It must be dim and distant, and will certainly be difficult to find, but its gravity could possibly affect the Pioneer spacecraft now heading out of the Solar System; so it is just conceivable that we might be on the threshold of a great astronomical discovery.

Comets and meteors

Comets are members of the Solar System and, rather surprisingly, there must be many millions of them, even though only about 10 to 20 might be visible in a year. Most of the time they are very distant from the Sun and too faint to be seen through even the largest telescopes. Occasionally, and usually quite unexpectedly, one may appear from more or less any direction, grow larger and brighter and perhaps spread a spectacular tail across the sky as it rounds the Sun, before dwindling and fading away into the distance.

All cometary orbits are very elongated, and some of the periods may well be thousands of years — or even millions for those that reach out half-way to the nearest stars. A number of much shorter periods are known, but of these the only comet which is usually quite conspicuous is the famous Comet Halley, which has a period of 76 years, and only travels out beyond the orbit of Neptune.

Comets seem to consist mainly of ice and dust, and are frequently described as 'dirty snowballs'. At great distances from the Sun they are just tiny frozen bodies, perhaps about 10km or so across in size. As a comet approaches and is heated by solar radiation, gases and some dust are given off and appear as the head, or coma. The same materials also form the tail, which may sometimes have two distinct forms, with the ionized gases being controlled by the particles of the solar wind, while the electrically neutral gases and dust are dispersed along the comet's orbit.

Comet Halley's return in 1066 is shown in this scene from the Bayeux Tapestry.

Left: **a close-up of the head of Comet Halley in 1910.**

Some tails may be enormous, such as that of the Great Comet of 1811, which was greater than the distance between Mars and the Sun. On some occasions more than one tail can be seen — one comet had as many as seven — but on others the comet may appear as little more than a fuzzy patch. Sometimes a bright star-like nucleus is visible, and very occasionally this may split into two or more parts. It is probably this process which has given rise to the families of comets that are known, including those called Sun-grazers, which pass well within the solar corona.

Comets gradually waste away with each passage around the Sun, and eventually all that remains is a cloud of dust particles

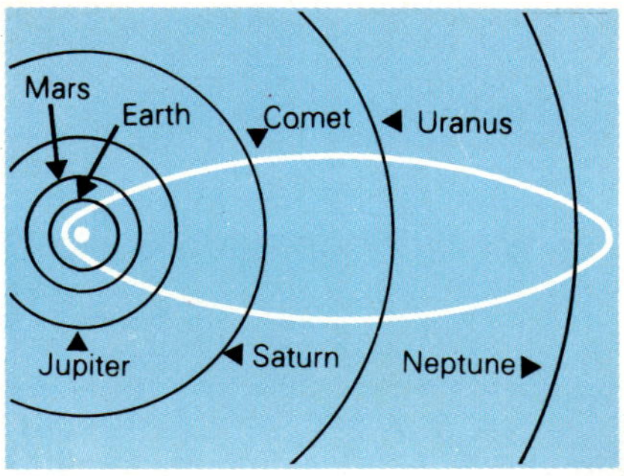

The orbit of Comet Halley extends out beyond the orbit of Neptune and has a period of 76 years.

The spectacular Comet Ikeya-Seki photographed in 1965.

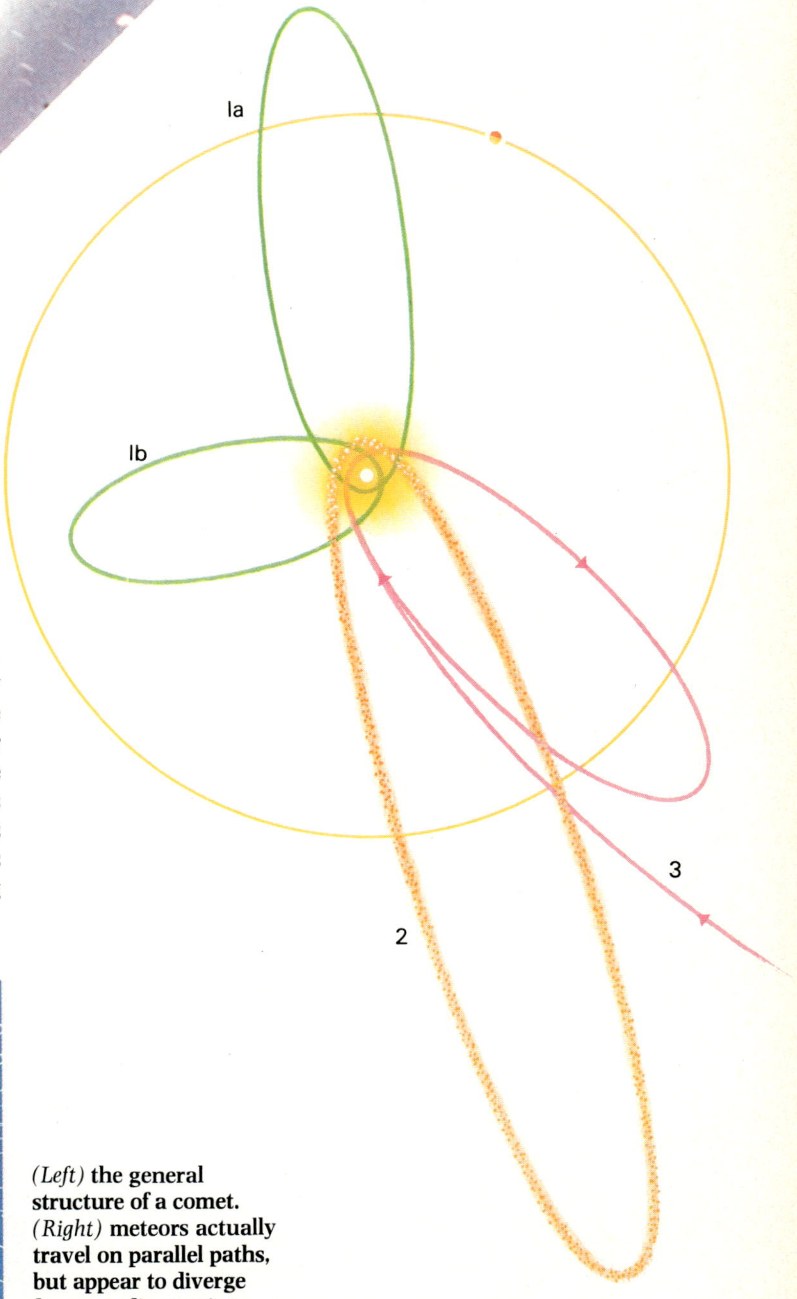

Short-period cometary orbits (1a and 1b), a meteor stream (2), and the capture of a comet into a smaller orbit (3).

spread along the orbit. These frequently enter the Earth's atmosphere, when they are fleetingly seen as they burn up as **meteors**, or 'shooting stars'. Although some are visible on any clear night, there are several periods during the year when considerable numbers are observed. Many of these meteor showers are linked with individual comets, including some that can no longer be found.

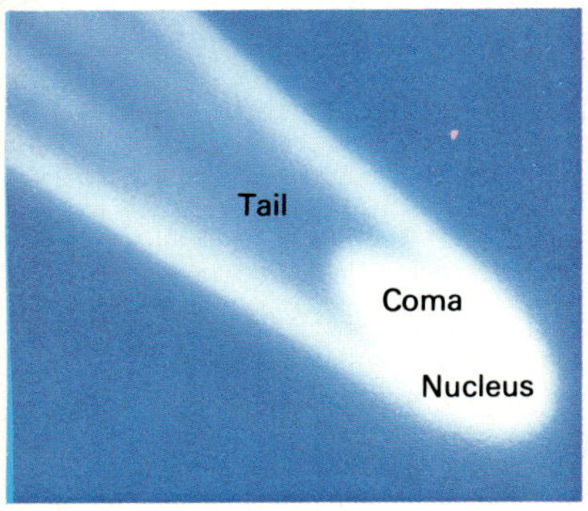

***(Left)* the general structure of a comet. *(Right)* meteors actually travel on parallel paths, but appear to diverge from a radiant point.**

Surveying the stars

To a casual observer looking up into the sky, the stars appear to be of different brightness (or of different **magnitudes,** to use the proper astronomical term). It is natural to wonder whether they really do have differing magnitudes, or whether they are all the same, but at different distances. And if it comes to that, what *are* their distances, and their sizes, temperatures, masses and compositions? The answers to all these questions can be given, but even today some are very much easier to answer than others.

Although very sophisticated methods are applied nowadays, it is essentially the colours which reveal the temperatures of the stars. Those that are red have the lowest temperatures (about 3000°C), and from this they range upwards through yellow stars like the Sun (6000°C) to brilliant blue-white stars of about 40 000°C. These temperatures are those of the stellar surface, rather than of the deep interior which is always much hotter still.

A major breakthrough came when it was discovered how sunlight (or starlight) could be analyzed by means of a spectroscope. In this instrument the light is dispersed by a prism (or in modern times, by a diffraction grating) to give a spectrum. With proper equipment many thousands of dark lines can be seen crossing the bright spectrum, and each element present in the star gives rise to its own individual set of lines. Once these have been determined from laboratory experiments and calculations, we have the answer to another of our questions and know the composition of the stars. In addition we have a very powerful method of distinguishing between the different types of star.

Distances proved very difficult to determine, although in theory, due to the Earth's movement around the Sun, any star should show a regular slight alteration in its position against the background of still further stars. Many astronomers struggled for years to make the necessary measurements. When success finally came the distances determined were almost unbelievably large: the *nearest* star, Proxima Centauri, proves to be at a distance of 270 000 astronomical units (see page 00), or – expressing it in terms of the distance travelled by light in one year – 4·26 light-years. Even today the measurement of distances is very difficult indeed, and usually other methods have to be adopted for their calculation.

The most important information about stellar size and mass comes from binary systems, where two stars are in orbit about one another, and most especially from systems where the members eclipse one another (see page 00). The range of sizes proves to be very wide, from tiny white dwarf stars about one-hundredth the size of the Sun, to red supergiant stars with diameters as much as 1000 times that of the Sun. But the range of masses is not so great – from about 0·06 to 60 solar masses overall.

Above: **this view towards the centre of the Galaxy gives some indication of the vast number of stars which form our system.**

Right: **examination of the lines in a spectrum indicates the various elements present in the atmosphere of a star.**

In the classification of stars, the temperatures and sizes are strongly dependent upon their masses.

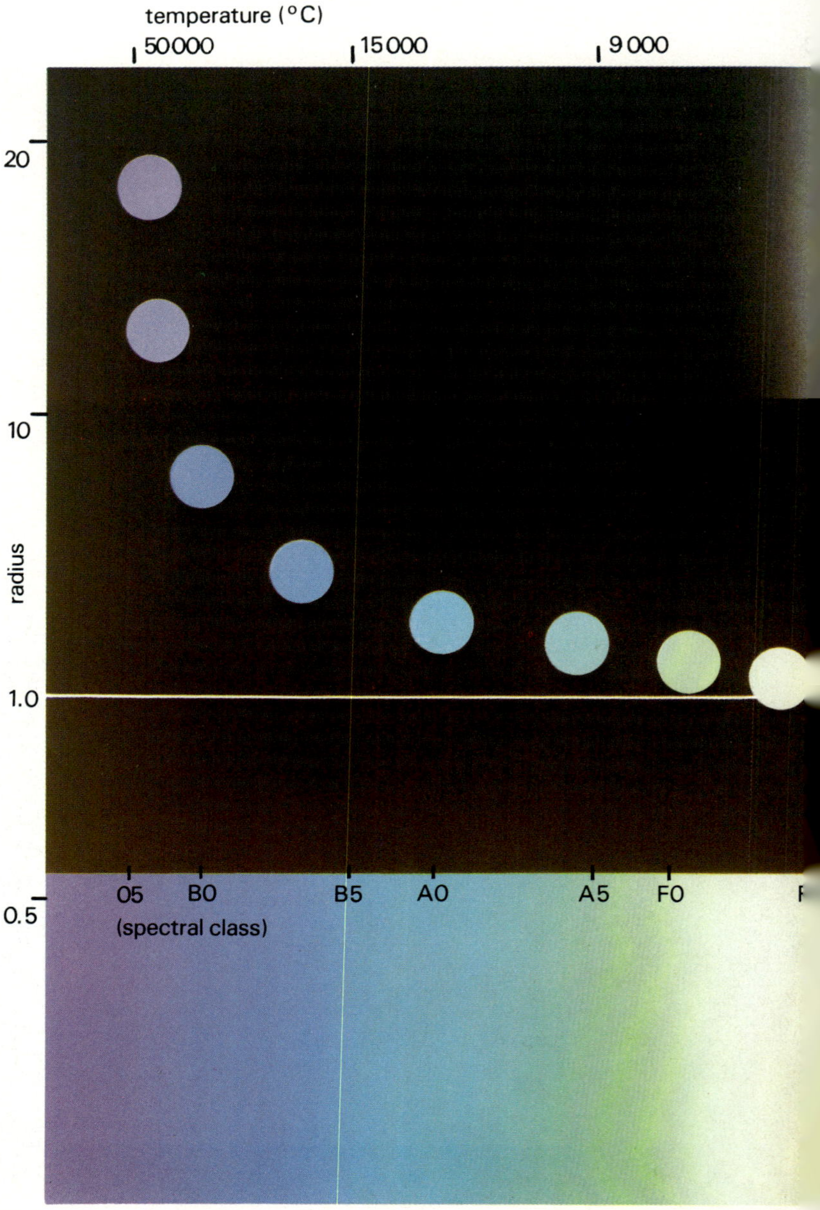

Both Sirius, the brightest star (5) and Procyon (1) have white dwarf companions. Aldebaran (2) is a giant, while Betelgeuse (3), Rigel (4) and Arneb (6) are supergiants.

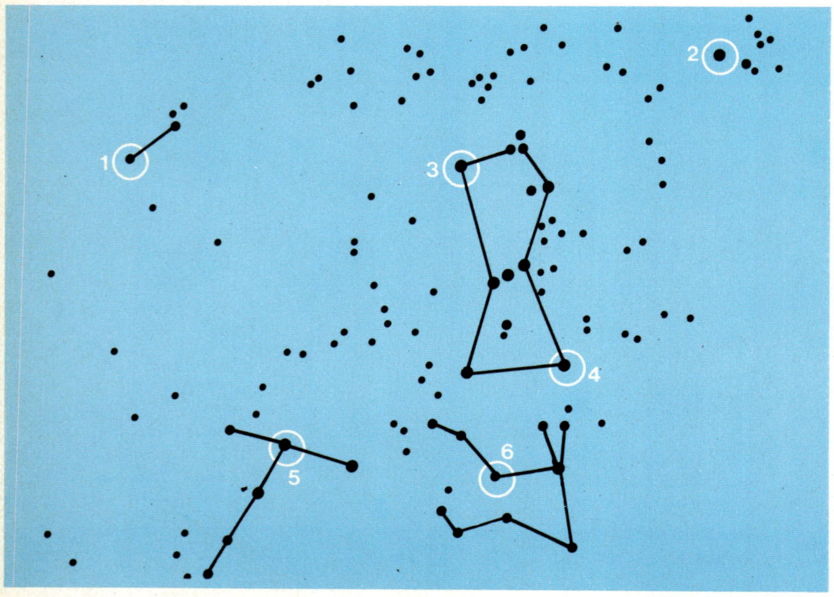

Below: **this photograph of Orion shows Betelgeuse (1) and Rigel (2) as well as the 'belt' and 'sword' and the famous Orion Nebula.**

Far right: **some of the vast range of stellar sizes is shown in this diagram, ranging from a red supergiant down to a tiny white dwarf.**

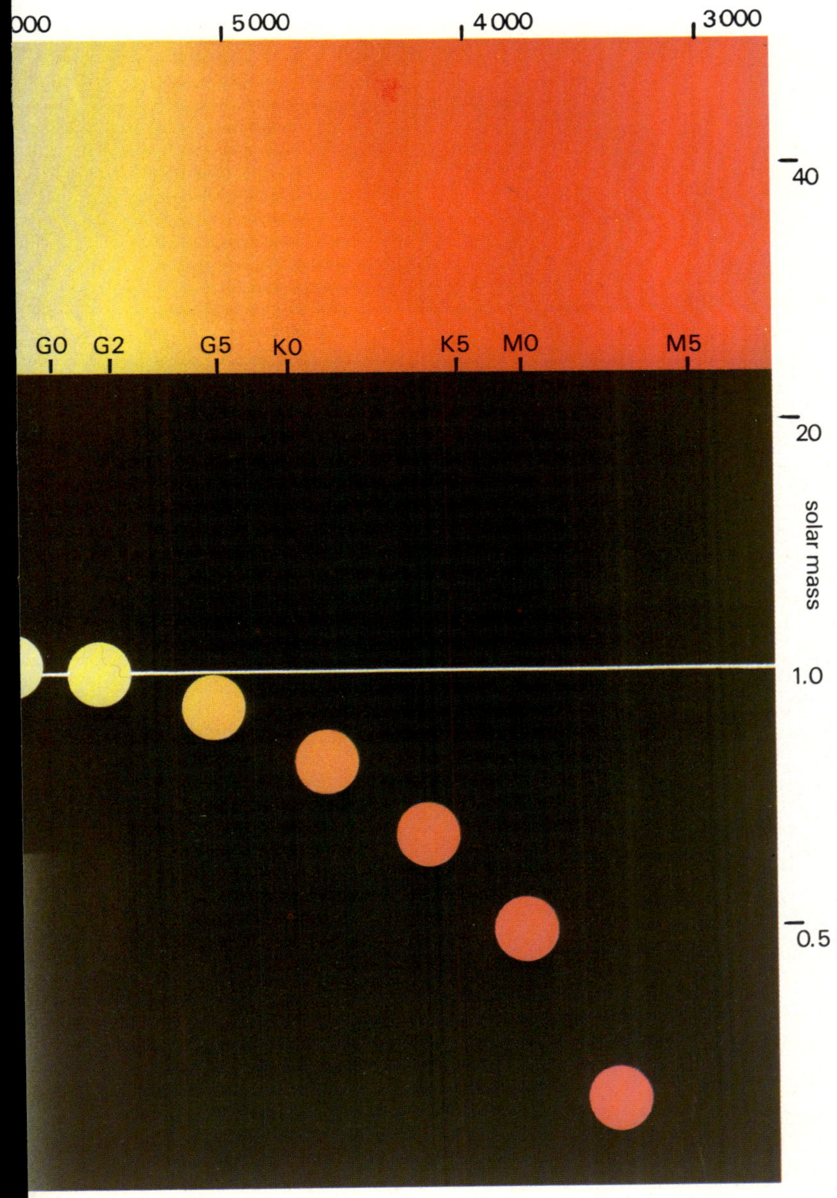

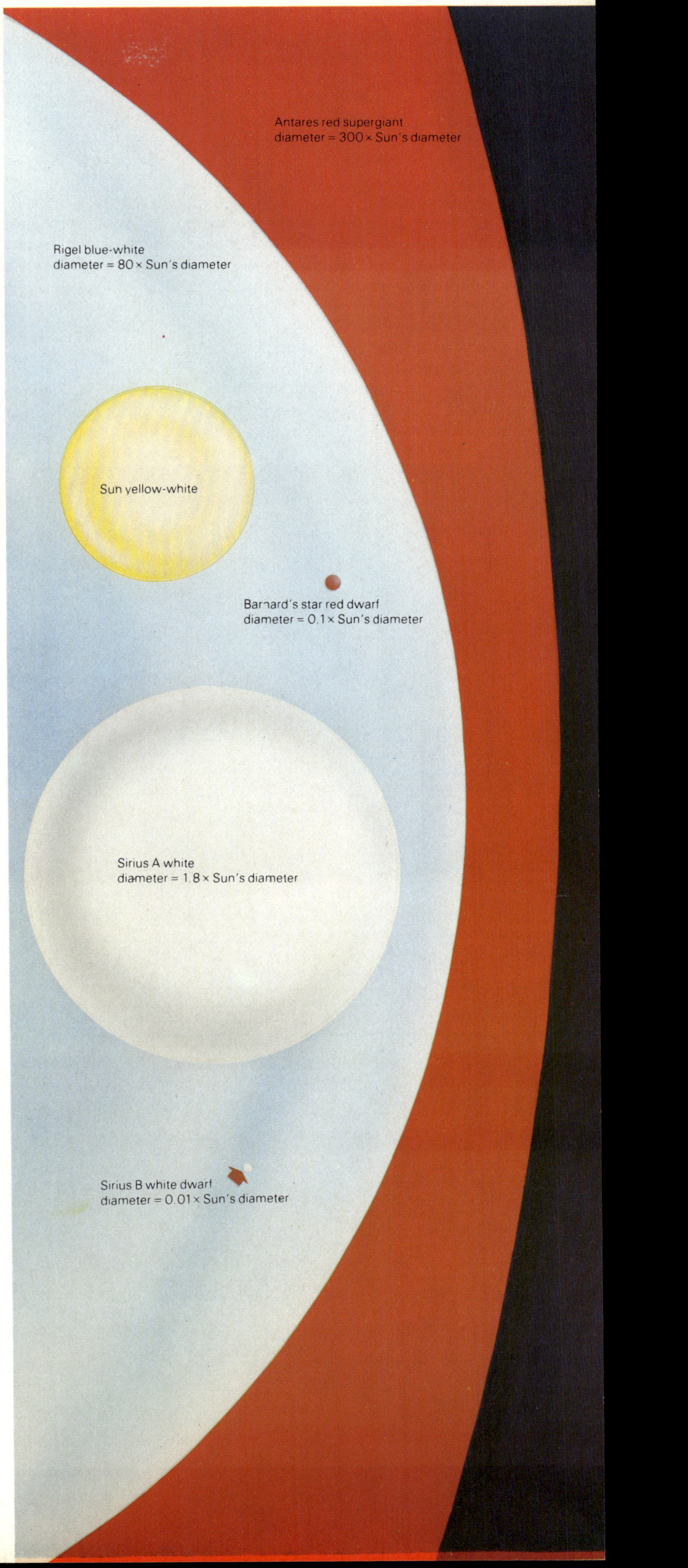

Stellar systems and clusters

Very few stars occur alone in space, as the Sun does. The grouping of stars into the various constellations, of course, very rarely has any physical reality. In most cases – and this happens also with some close visual double stars – they just happen to be close to the same line of sight, and are actually at very different distances. All stars have their own individual motions within the Galaxy, and in some cases this may be detected by making observations over a number of years. Over a long period of time – many thousands of years – the constellations will become distorted and eventually quite unrecognizable.

The vast majority of stars have been found to exist in true **binary** or **multiple systems**, where the stars are gravitationally linked. In these cases the stars are in orbit about one another – or strictly speaking, about the common centre of mass. As so many systems occur, it is natural to wonder why the Sun has no companion, or whether in fact it does have one which is too faint and distant for us to see.

In many systems the stars are too close for them to be observed separately, and it is only by spectroscopic analysis that we may determine that they are doubles. This is usually the case with eclipsing binaries, where the orbits are aligned with our line of sight, and changes of brightness occur as the stars pass in front of one another. The range of orbital periods is very great: some of the visual binaries have periods of hundreds of years, while some eclipsing and spectroscopic systems have periods of less than 2 hours.

In many areas of the sky there are obvious **clusters** of stars – the Pleiades, in the constellation of Taurus, are a good example. When these are examined in detail the stars are found to be very young, usually only a few million years old, so we may assume that they were all born at the same time. This is supported by modern theories which suggest that stars begin their lives within large dust and gas clouds, and that instead of stars being born singly, several are formed together. With the passing of time such loose clusters are likely to become less distinct, as the stars gradually disperse due to their own individual motions.

Quite apart from the loose (or galactic) clusters, there are also very large spherical groupings of stars. These **globular clusters** may consist of hundreds of thousands of stars and their ages are very great. They date back to an early stage in the history of the Galaxy, and their stars show a simpler, more primitive chemical composition than the Sun and similar stars.

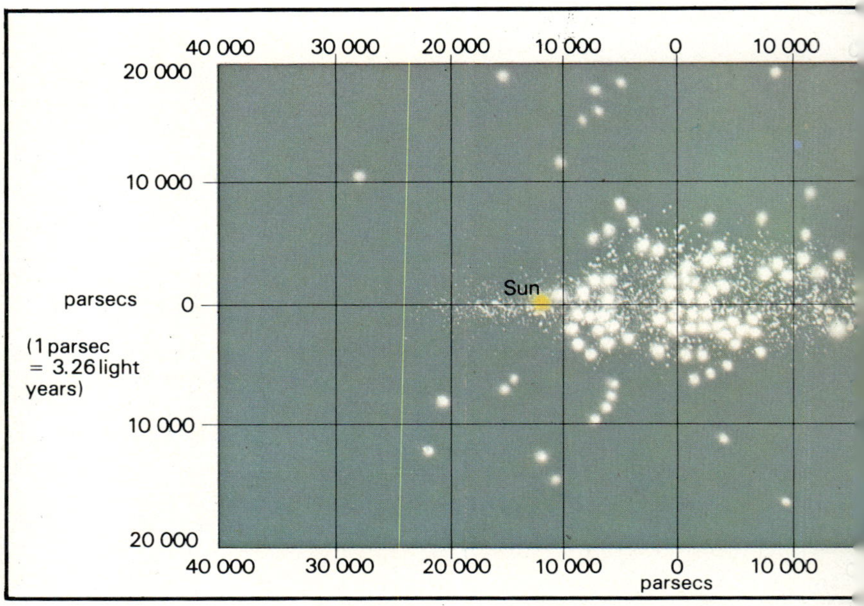

The stars in the familiar constellation of Cassiopeia are actually at very different distances from us.

The Pleiades form part of a very young, open cluster. Some of the bright blue stars are surrounded by nebulosity.

Above: **the distribution of globular clusters in our Galaxy.**

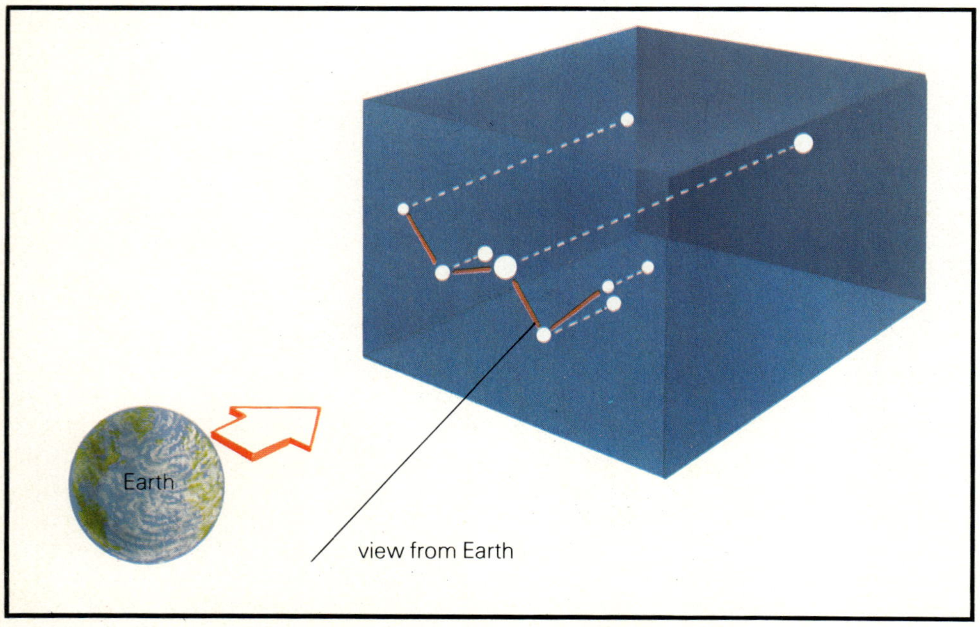

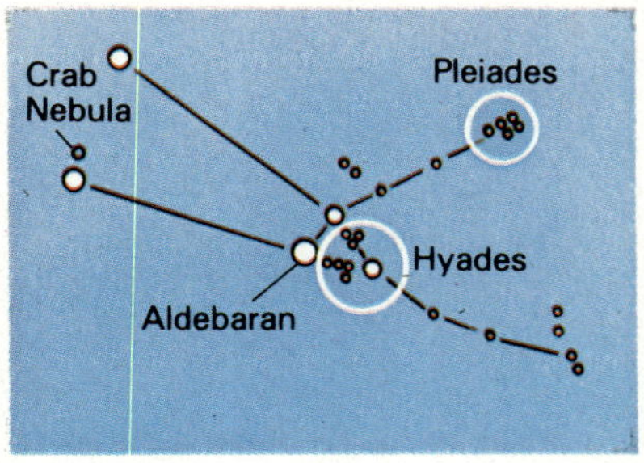

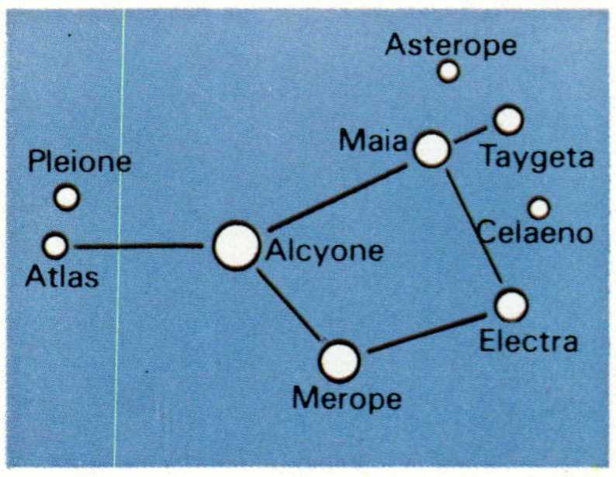

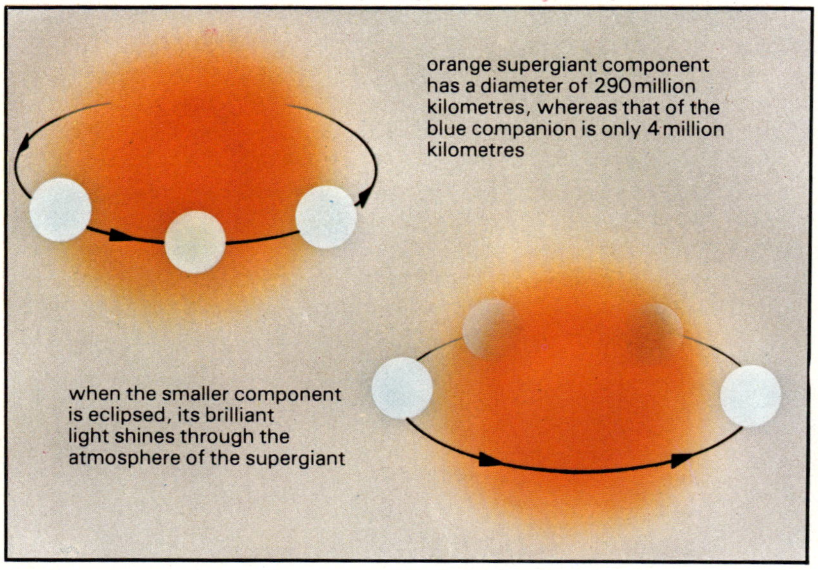

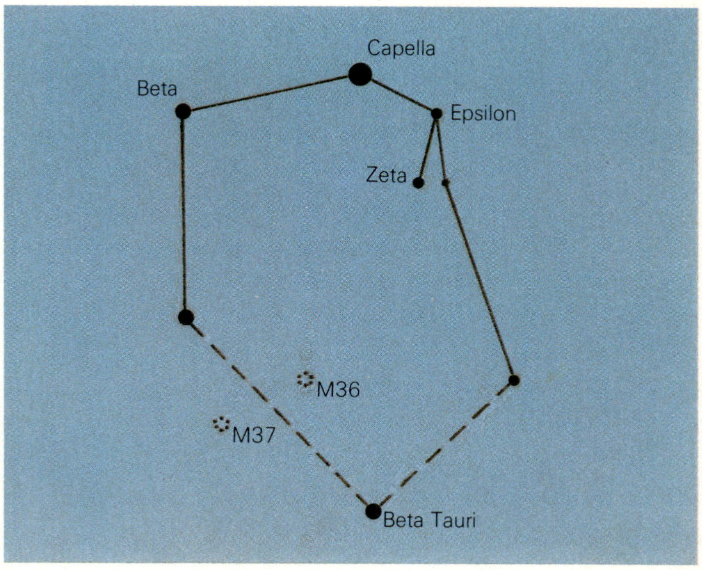

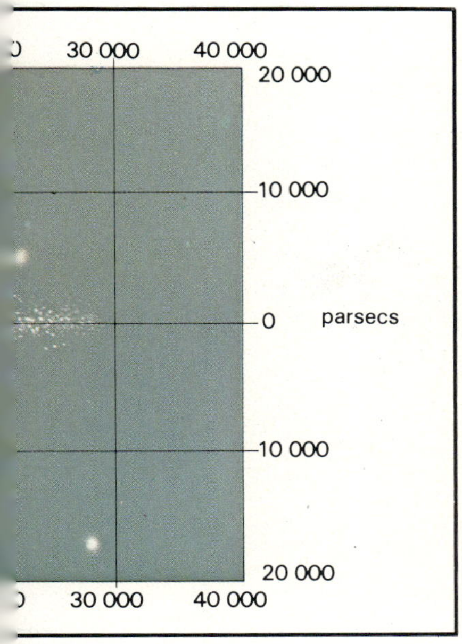

Auriga, showing the positions of the famous eclipsing systems, Epsilon and Zeta.

Above left: primary eclipses in the Zeta Aurigae system occur every 972 days.

Centre: the globular cluster M13 in Hercules consists of many thousands of stars.

Left, top: the location of the Pleiades and Hyades clusters. *(Bottom)* The brightest stars in the Pleiades.

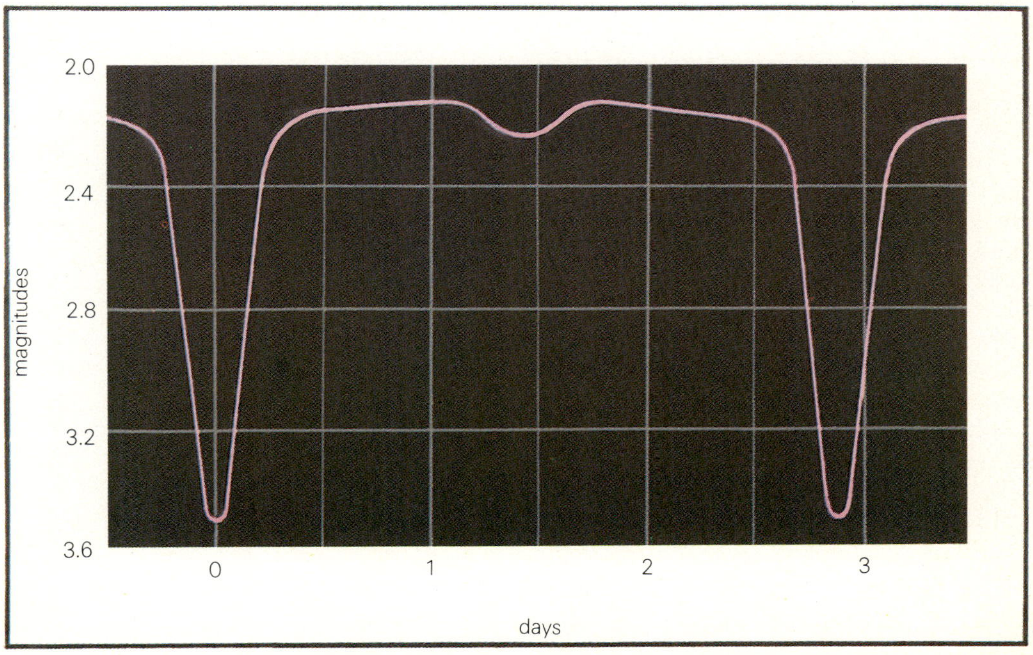

The light-curve of the famous eclipsing binary Algol (Beta Persei) shows a regular variation in brightness.

Nebulae

Even with the naked eye it is possible to discern amongst the stars in the sky (which appear as points of light) a few hazy spots. These cloud-like patches, or **nebulae**, attracted the attention of some early telescopic observers, and gradually more examples were found and their details recorded. The most famous catalogue was that compiled by the French astronomer Charles Messier, to help prevent confusion in his comet searches, which was finally published in 1784. The reference numbers from this catalogue remain in use to the present day. Extensive work in this field was later carried out by Sir William Herschel and his son, Sir John. Some of the objects noted have since proved to be mere chance groupings of stars, while others are definite open or globular clusters. Yet others – later termed the 'spiral nebulae' when their structure began to be visible in improved telescopes – we know to be distant galaxies. There remained some objects which could not be resolved into individual stars: the **planetary nebulae**, so named from their resemblance to planetary disks, and irregular objects with little apparent form, of which the great Orion Nebula is an excellent example.

Only when the spectroscope became available as an auxiliary instrument fitted to the telescope could it be determined that these last two classes of nebulae actually consisted of glowing gas. Even then some time was to elapse before it was finally established that the gas was ionized by the radiation from neighbouring stars. In the case of the planetary nebulae, which are shells of gas shed by evolving stars, the stellar remnant responsible for the excitation is visible in the centre, but in many of the other objects the bright ionizing star (or stars) may be hidden.

In certain parts of the sky – the Great Rift in Cygnus for example – there appear to be large dark patches with either no stars at all, or at most with very, very few. Sir William Herschel believed these to be true 'starless voids in space', but subsequent studies showed that they were, in fact, filled with dark obscuring matter. This dust is particularly concentrated in the plane of the disk of the Galaxy, but also occurs in association with many of the glowing gaseous nebulae. These dense dust clouds are frequently detectable by the infrared radiation from dust grains heated by unseen stars. Observa-

The Rosette Nebula. Silhouetted against the glowing hydrogen gas are dark clouds where stars are probably being formed.

Right: **the Horsehead Nebula in Orion is just a small part of a vast dark cloud of dust.**

tions at microwave and other radio frequencies detect related vast clouds containing many complex molecules – including such substances as ammonia and alcohol – once thought to be non-existent in interstellar space, or anywhere other than on planets.

The densities within some of these massive clouds are so high that they are ideal sites for stellar formation. Many young stars have been detected within the region of the Orion Nebula in particular, as well as in other similar objects.

The North American Nebula, where the 'Bay of Mexico' is a cloud of obscuring dust.

Above: **the Veil Nebula is a remnant of a very ancient supernova explosion.**

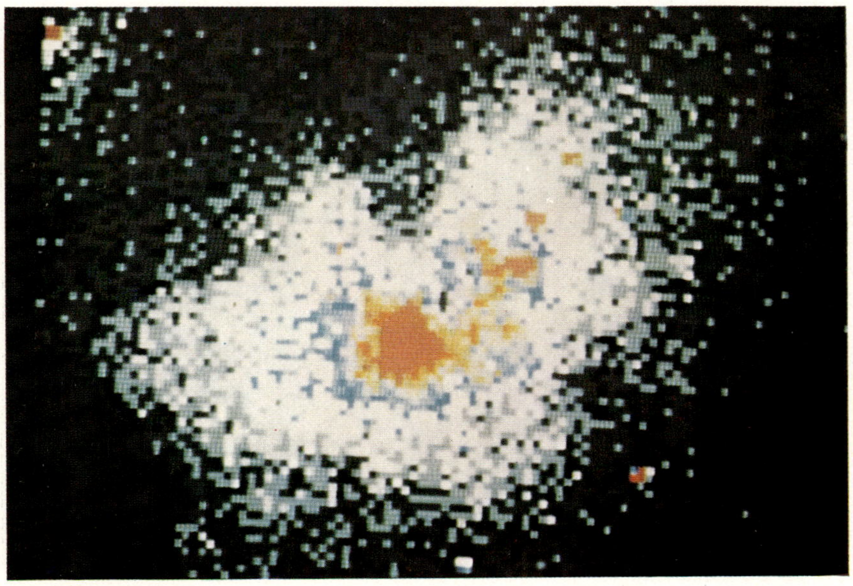

Below: **the great Orion Nebula as seen in ultraviolet light.**

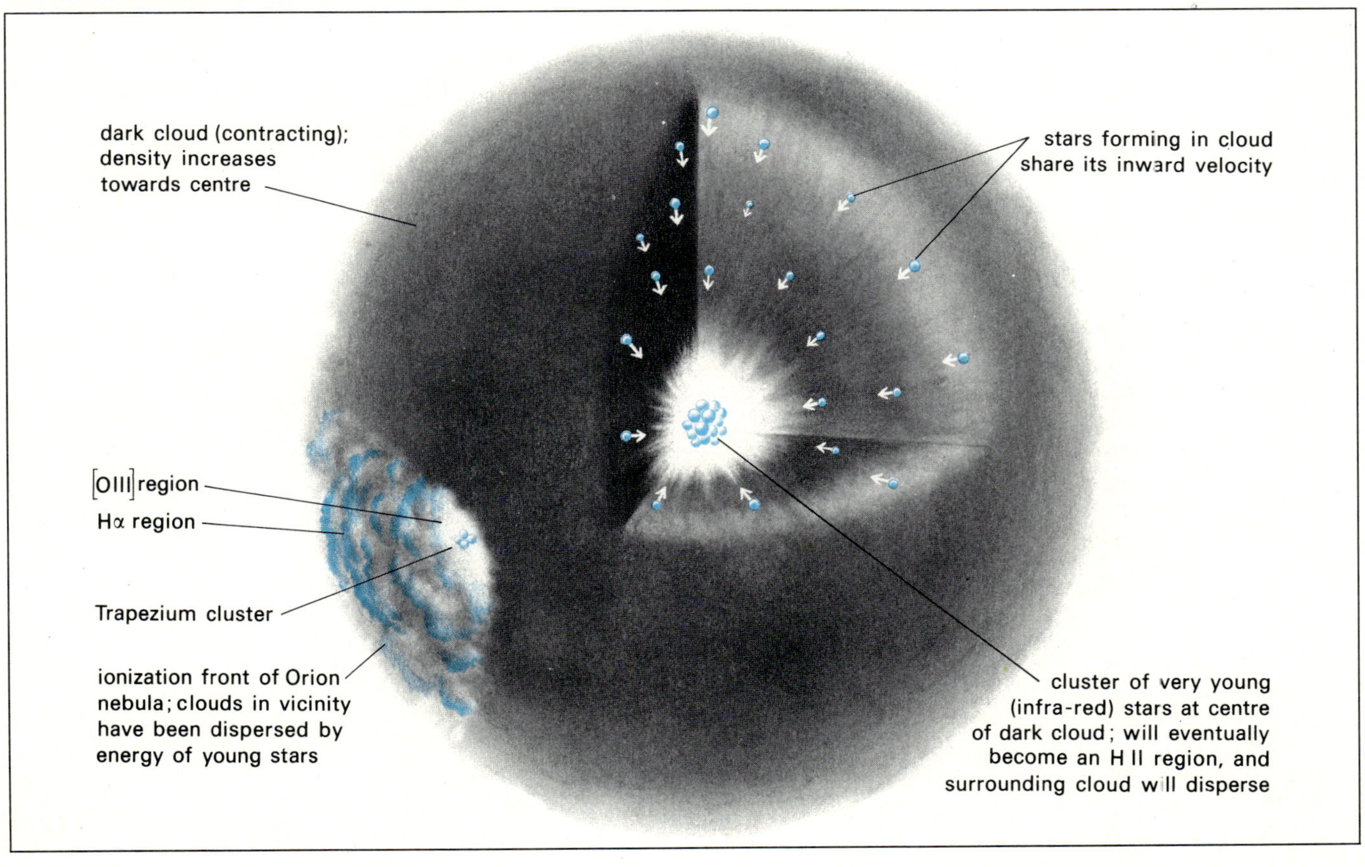

The Orion Nebula is just a tiny glowing part of a vast interstellar cloud of dust and gas.

How stars evolve

The giant clouds of gas and dust within our Galaxy are sites of stellar formation. How this is initiated is uncertain, but it may be due to compressional waves which travel around the Galaxy, or perhaps to shock-waves from supernova explosions. Once a certain degree of compression has taken place, gravity takes over and continues the process. The **proto-star** begins to heat up and may be detectable, even through its cocoon of gas and dust, by the infra-red radiation which it emits. Eventually any protostar will become so dense and hot that nuclear fusion starts in its interior, and it begins its independent existence as a true star. After an initial unstable period it will settle down at a size, temperature and brightness determined by the amount of material within it.

Every star spends most of its life quietly converting hydrogen to helium within its core; the Sun will do this for about 10 thousand million years, although more massive stars have much shorter lives. When a lot of the hydrogen is exhausted, helium begins to burn to carbon. The star then expands greatly, becoming a **red giant**; when this happens in the case of the Sun it will be about 100 times its present size – about 130 million km across, nearly filling the Earth's orbit. At a slightly later stage the star gently sheds its outer layers which expand out into space around it, giving rise to a planetary nebula. The core remains and contracts to a **white dwarf**, slowly radiating away its remaining energy, until it becomes an extinct **black dwarf** star.

More massive stars evolve in a different way, and they may burn carbon into yet heavier elements, such as iron, neon and magnesium. The very largest stars may even

Opposite top left: **the nuclear fusion reactions whereby protons (orange) are converted, via deuterium and helium-3, to stable helium-4, with the release of energy. (Neutrons are in purple.)**

Opposite top right: **as well as a protostar planetary bodies may also form (a). In its early stages the star may lose much material in a strong stellar wind (b). At a much later stage it may expand into a red giant.**

Below: **bright young stars are frequently found in regions rich in dust and gas.**

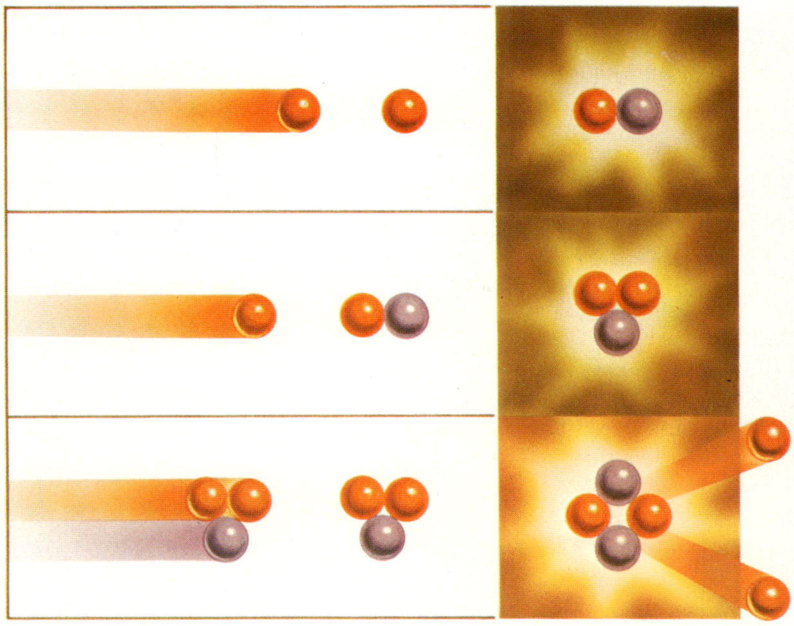

Below: the Ring Nebula in Lyra is a shell of gas shed by an evolving star.

Bottom: the vast interstellar clouds (a) are the sites of stellar formation. Individual regions begin to contract under their own gravity (b).

Right: this diagram shows the path (beginning centre right) followed by stars like the Sun during their lifetime.

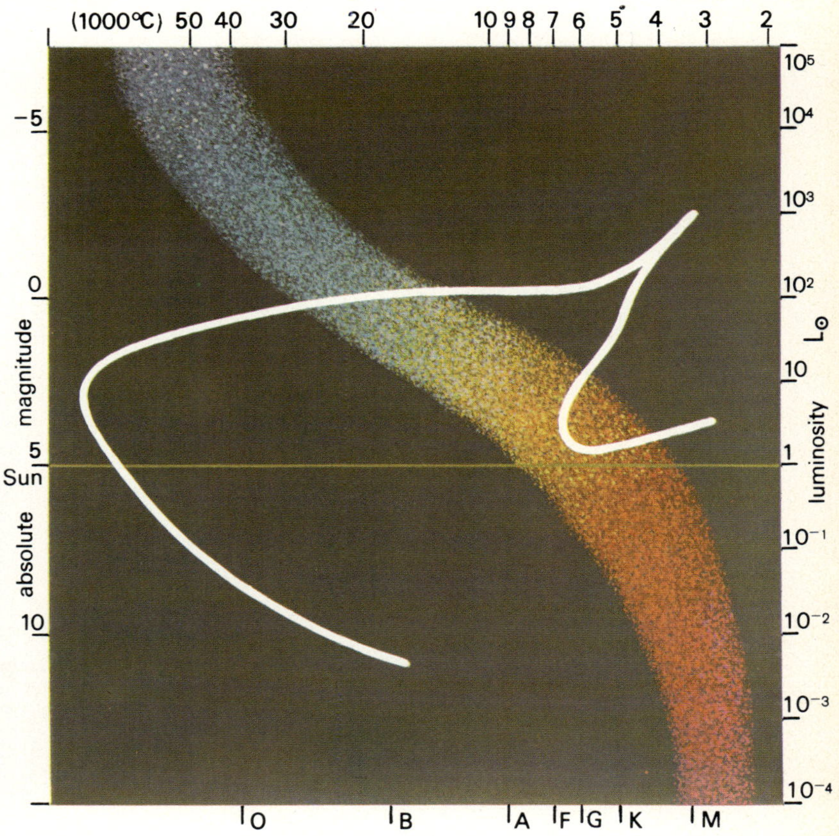

turn these into elements as heavy as iron, beyond which the process cannot continue. The most massive stars spend most of their short lives as brilliant hot **supergiants,** before expanding into the truly enormous **red supergiant** stars.

All stars go through stages when they pulsate in size, expanding and contracting, with associated changes in brightness. Particularly important are the supergiant **Cepheid variables,** where the pulsations and light changes are very regular, and rigidly dependent upon the mass. If a star's period can be determined then its true brightness – its *absolute* magnitude – follows, which by comparison with its *apparent* magnitude will give its exact distance. These stars are so bright that they can be used to determine the distances of other galaxies.

Exploding stars

Sometimes the appearance of a familiar constellation is completely changed when a brilliant star suddenly becomes visible where none could be seen before. Slowly, but inevitably, the star fades and may become invisible even to the largest telescopes. Such spectacular changes have always been noted by astronomers, the earliest known occurrences appearing in Chinese records dating back about three thousand years. Although we now know that they are not 'new' stars the term **nova** is still used to identify them.

Novae are not single stars, but occur in binary systems. One star, originally the more massive, evolves faster than the other, goes through its giant phase and becomes a white dwarf. The second star, which is itself just beginning to evolve and expand, is no longer able to retain some of the material in its outer layers, and this hydrogen gas spirals down on to the white dwarf. Eventually, when enough gas has accumulated, it suddenly ignites in a thermonuclear explosion. Some material is thrown off and the resulting small nebulosity may be detectable with very large telescopes. This may happen several times at intervals of a few thousand years.

Bright explosions of this general sort had been seen both in the so-called 'spiral nebulae' and, as we now realise, in our own Galaxy. When it was recognised that the 'spiral nebulae' were actually separate distant galaxies, it became obvious that these stars were far more brilliant than ordinary novae, and they thus became known as **supernovae.** A supernova sometimes outshines the whole of the galaxy in which it occurs – a galaxy that may contain ten thousand million stars!

It is now believed that there are two types of supernovae, and that one closely resembles the novae just described: a white dwarf in a binary system gains material from its companion, but there is a very precise limit to the mass which such a highly evolved old star may have, and when that limit is exceeded an explosion completely disrupts the star.

In the second type of supernova, a single star, greater than ten times the mass of the Sun, has burnt the elements within it until iron has formed. No further energy-producing process is available, so there is nothing to sustain it against the force of gravity, and a collapse results. As it does so the outer layers are violently expelled into space, while the core becomes a tiny and incredibly dense body known as a neutron star, probably only 30km or less in diameter. When these objects emit rapidly pulsating radio waves they may be detected as **pulsars.**

Right: **the Crab Nebula is the remains of a supernova explosion observed in 1054 by Chinese and Arab astronomers.**

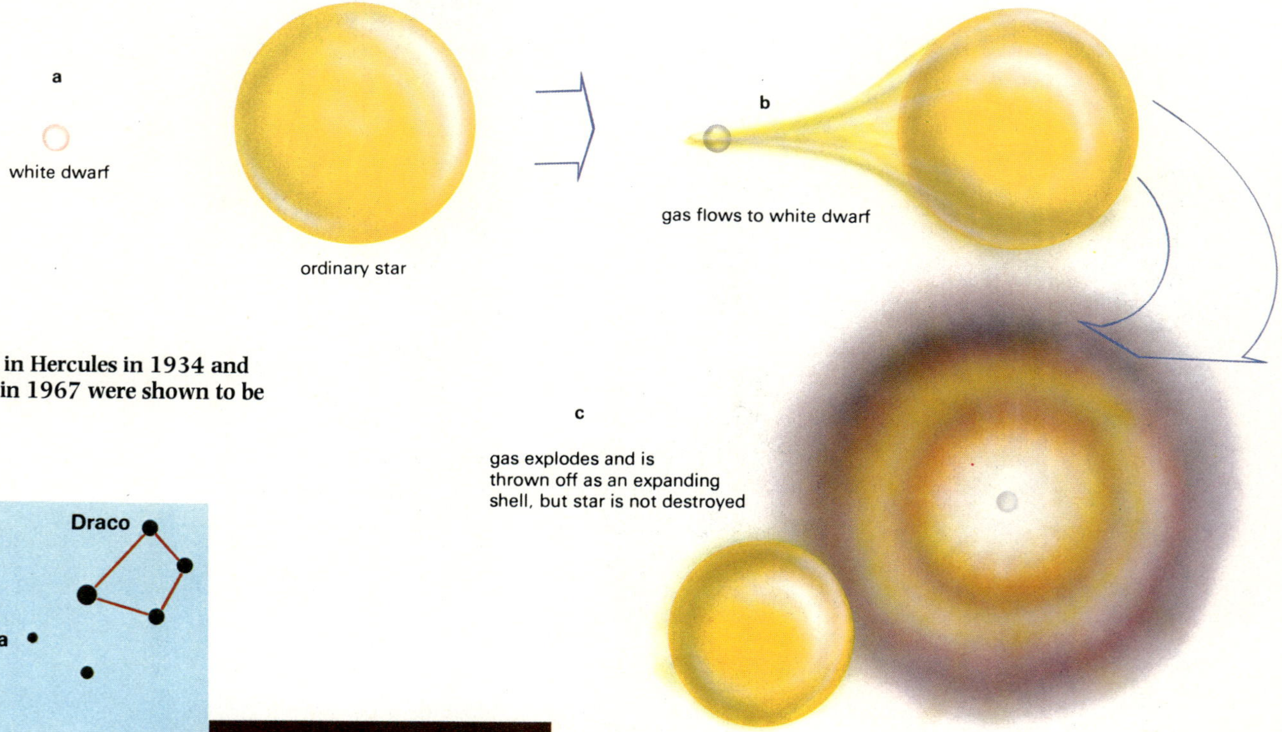

Above: **a nova can actually explode many times, at intervals of thousands of years, as the amount of gas which ignites is only very small.**

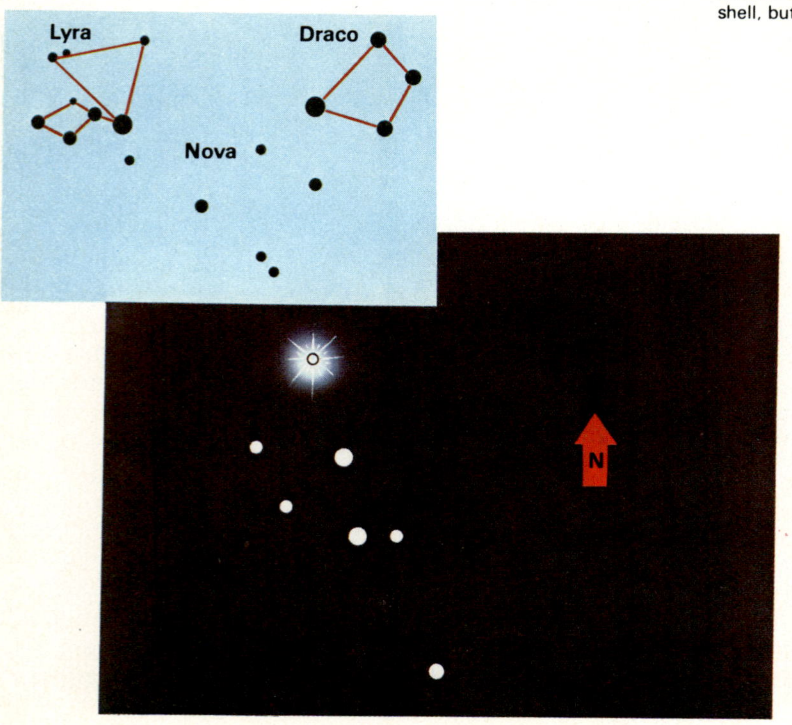

Below: **both the nova in Hercules in 1934 and the one in Delphinus in 1967 were shown to be binary systems.**

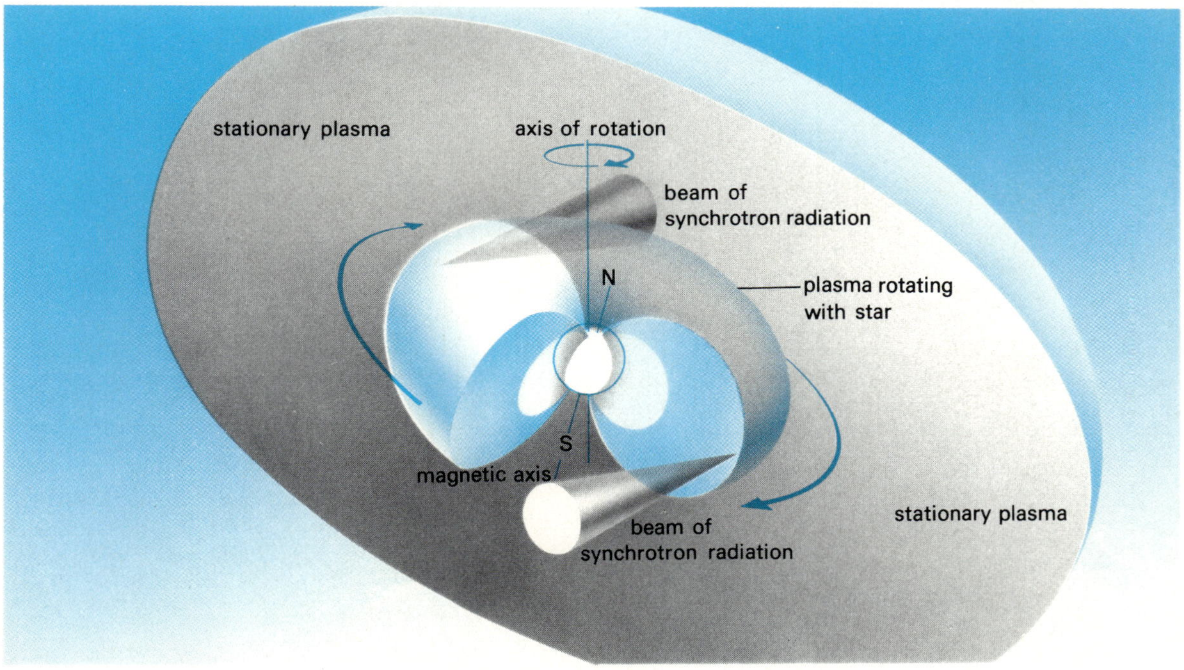

Left: pulsars produce their bursts of radiation by a form of 'lighthouse effect'.

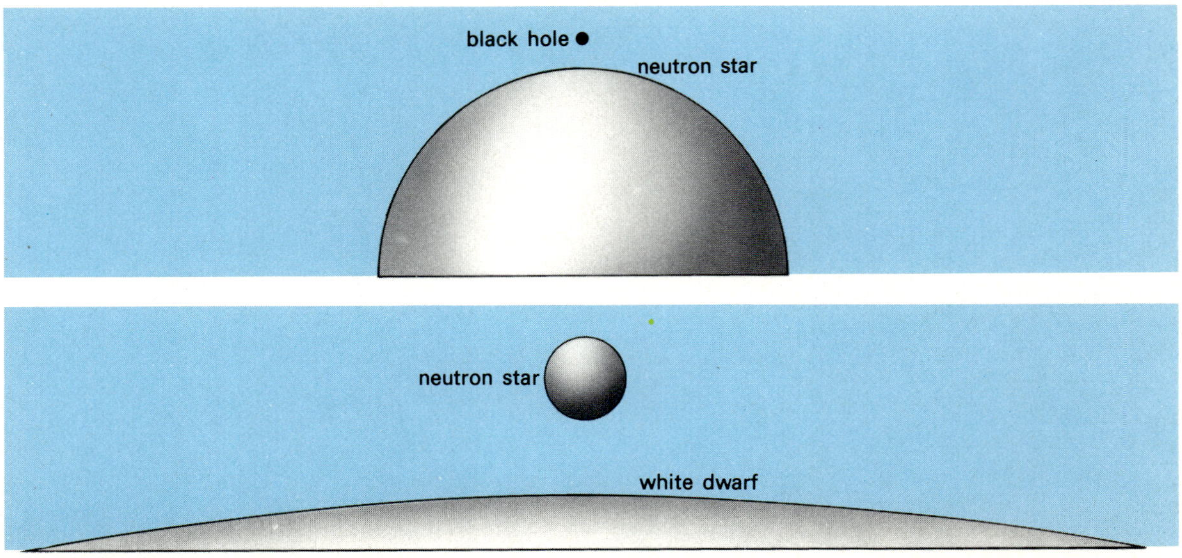

A comparison of the sizes of a white dwarf star, a neutron star and a black hole, all of the same mass.

Black holes

Some of the objects which remain as the results of advanced stellar evolution are very strange. **White dwarf** stars, for example, which is what most stars will become, may be no larger in size than the Earth, yet contain as much mass as the Sun itself. The internal pressure in these bodies is so great that the material at the centre is no longer a gas, despite the high temperature of about 1 000 000°C, but depending upon the exact composition may be either a liquid or a crystalline solid.

Neutron stars are even more exotic. This time the equivalent of the Sun's mass is compressed into a sphere no more than about 30km across – about one hundred-millionth of the volume of a white dwarf! Such a dramatic reduction in size comes about because on an atomic level the elements have not retained their identity. Instead of the normal structure of an atom, with a nucleus (consisting of protons and neutrons) surrounded by 'shells' of electrons at a relatively great distance, the electrons have been driven into the nucleus to combine with the protons. The result is that the whole stellar mass (with the exception of a thin outer shell) has become one 'sea' of anonymous neutrons.

Theoretical calculations indicate that neutron stars cannot exceed two to three solar masses. If a supernova explosion leaves behind a core of greater than this amount, no known forces can stop it from collapsing indefinitely. It will shrink (and continue shrinking) to an infinitely tiny point which is known as a **singularity.** The density and gravitational effects of such an object also become infinitely great at the very centre, but at a certain distance in towards the singularity the velocity required to escape from its gravitational field becomes greater than the speed of light. As nothing can exceed the speed of light, it is not possible for any material object to escape from that central region, and neither can light itself. The object has become a **black hole.** The maximum diameter of a black hole is just a few kilometres, and the greater the mass involved, the larger the resulting black hole will be.

Although a black hole is invisible, its gravitational effects remain, and may make it possible for it to be detected. Several times it has been suggested that only black holes could account for the properties of certain objects, and that they may exist, for example in the centre of galaxies and some globular clusters. This has not been proved, however, and to date the most likely candidates remain two binary systems which are known to contain massive collapsed objects for which no visible evidence can be found.

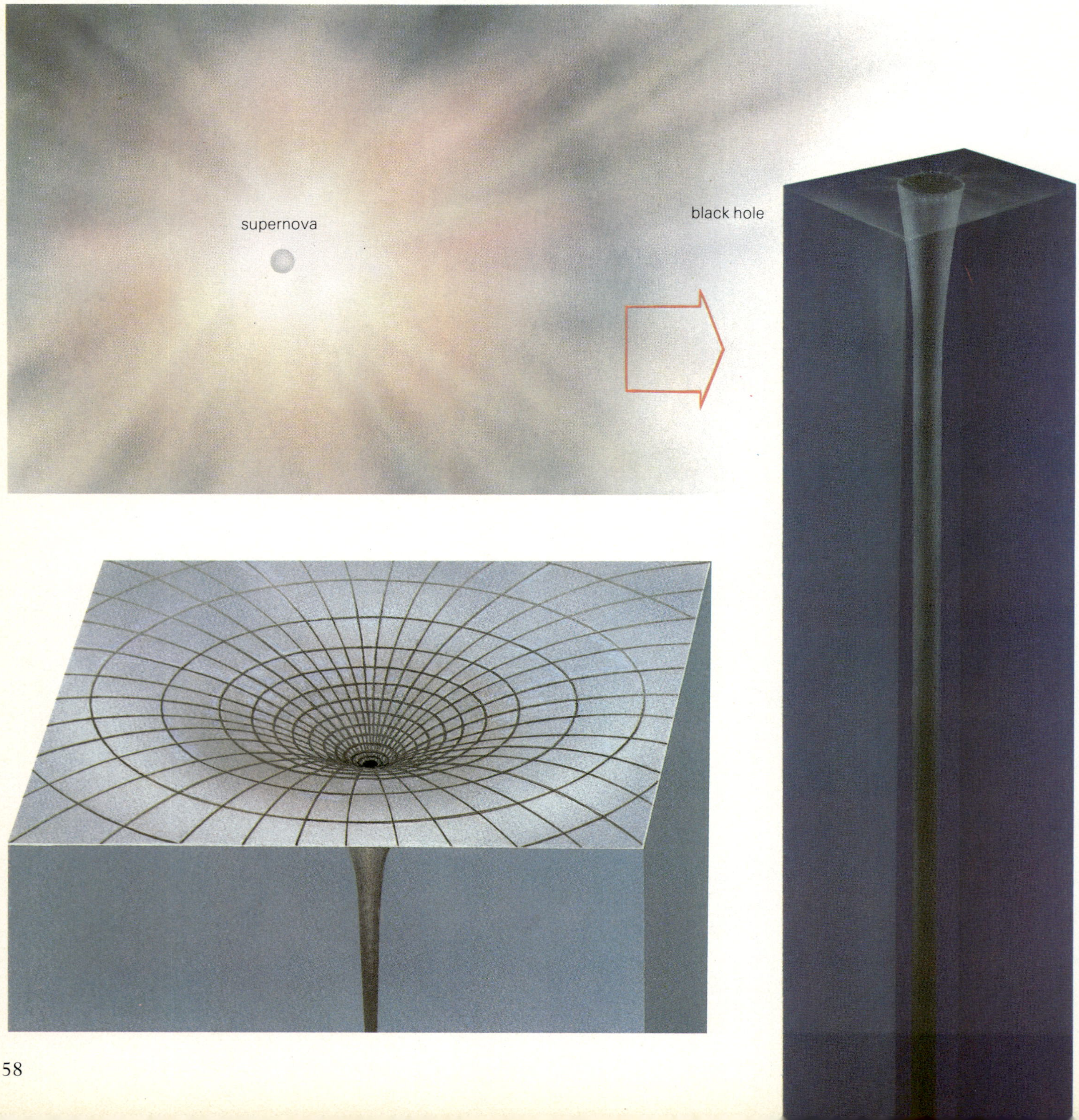

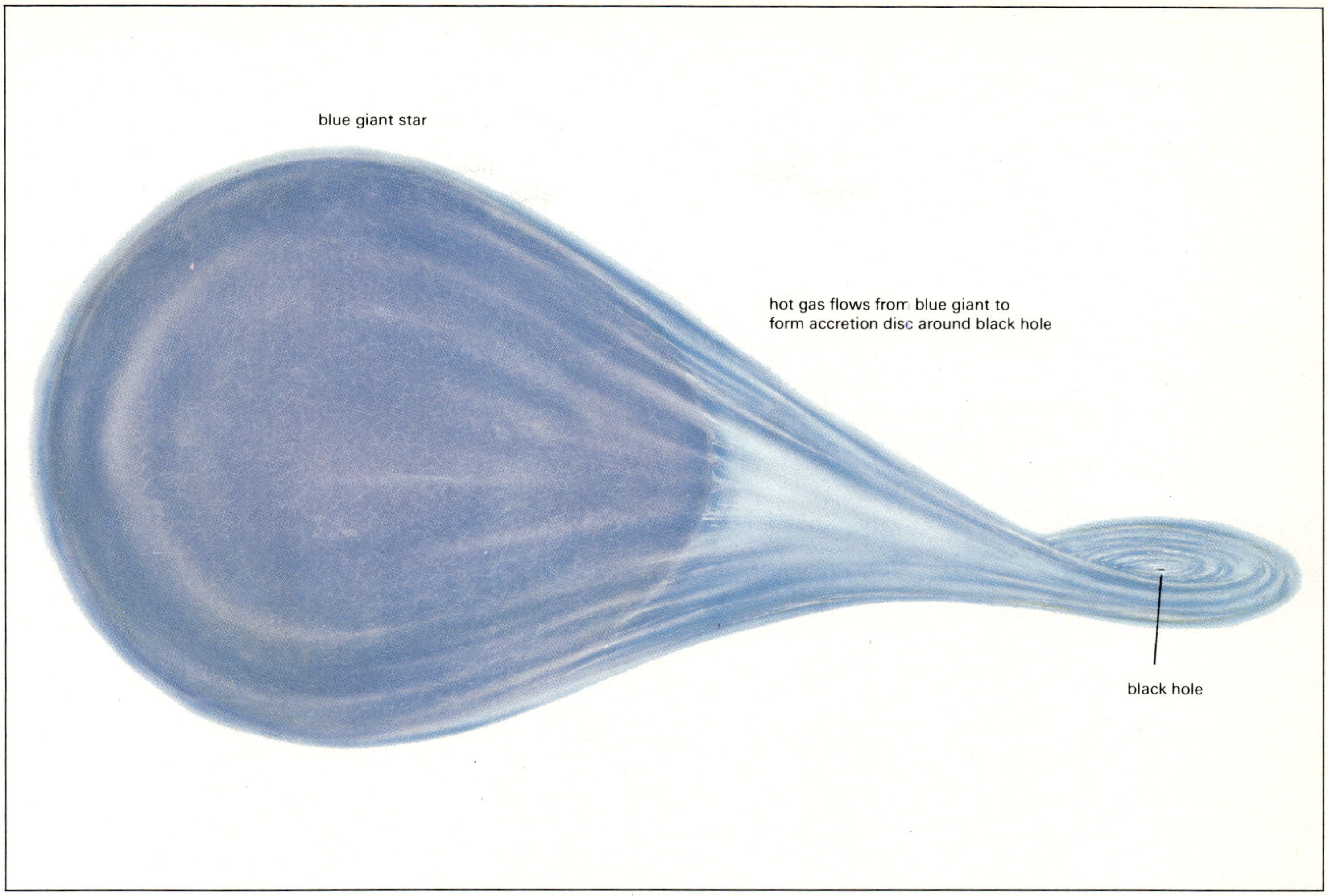

In Cygnus-X1 a large blue star appears to be losing material to an invisible black hole companion.

Opposite top: a black hole may be formed after a massive star explodes as a supernova.

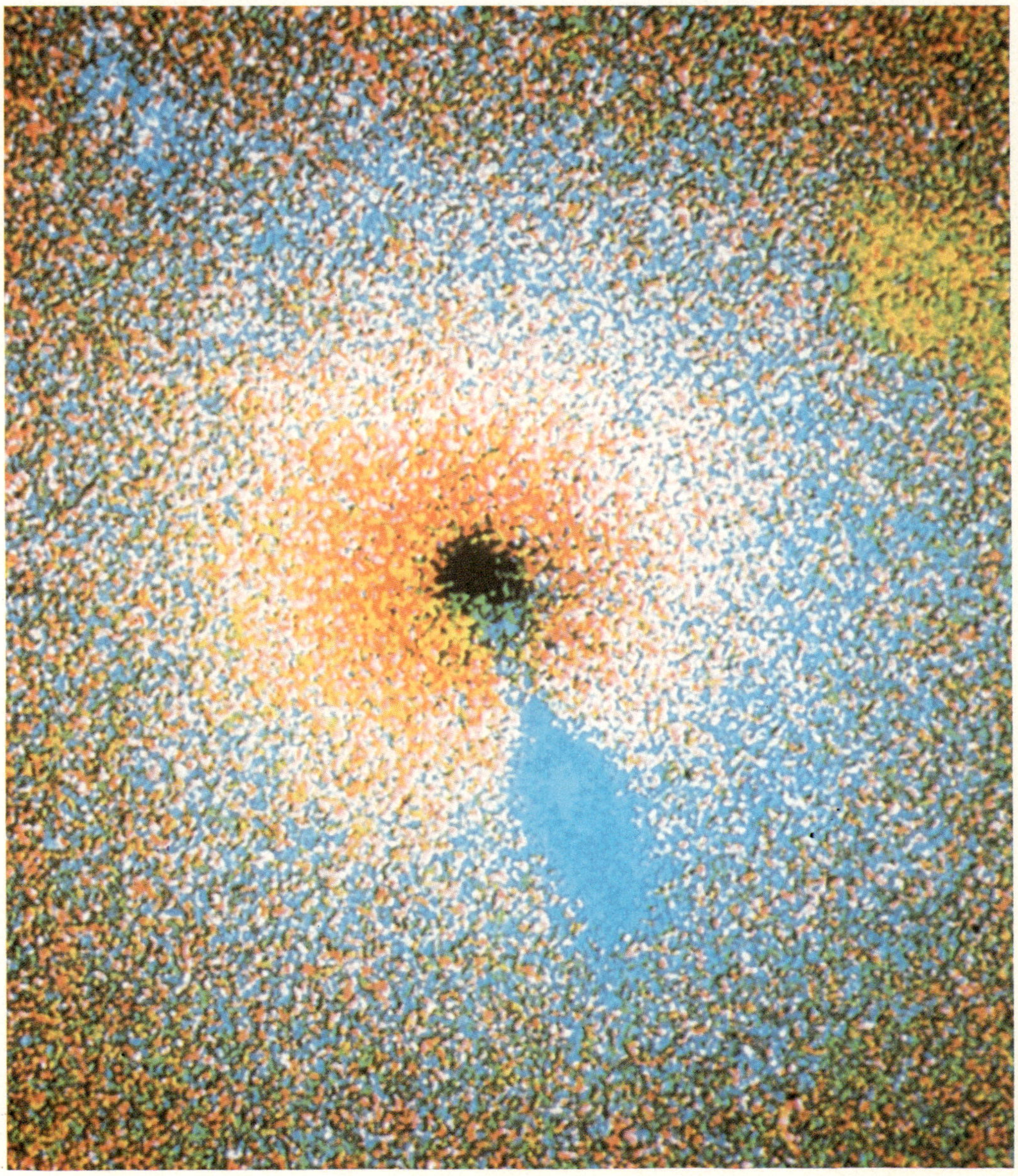

Opposite bottom: a black hole's gravity is so strong that the space around it is strongly distorted.

The giant galaxy M87 probably has a massive black hole at its centre.

Our Galaxy

One of the first discoveries made by Galileo Galilei when he turned his primitive telescope on to the sky was that the Milky Way, that hazy band of light stretching across the heavens, was formed of untold multitudes of stars. Yet that visible Milky Way is but a fraction of the vast system of stars which is the Galaxy inside which we live.

The Milky Way is all that we can see of a great disk of stars, some 100 000 light-years across, most of which is obscured by the great clouds of dust which lie within it. But infrared and radio waves can penetrate this screen to reveal that, like many other far distant stellar islands, our Galaxy also has long spiral arms of stars separated by comparatively empty lanes. The arms are marked by the great clouds of molecules, dust and gas, the bright nebulae, and the clusters of hot young stars. The Sun and Solar System lie on the edge of such an arm far out towards the edge of the great disk, the centre of which lies about 32 000 light-years away in the direction of Sagittarius.

Like those other distant stellar systems, our Galaxy also has a nucleus – a flattened ball of older, redder stars, about 20 000 light-years across and 9000 from pole to pole. Deep within this nucleus we find dense molecular clouds, a massive quantity of gas and, at the very centre, a tiny invisible object. Smaller than the size of Jupiter's orbit, yet containing the mass of several million Suns, this can only be a massive central black hole, probably formed very early in the life of the Galaxy. The galactic centre is very active with expanding clouds of material, much ionized gas, and strong sources of radio, X-ray and gamma-ray radiation. Compared with the violence which we see in some other galaxies, however, it is quiet and well-behaved.

Surrounding the disk and nucleus is an enormous, roughly spherical halo, the true extent of which is difficult to determine. It

The star clouds of Sagittarius hide the centre of our Galaxy from observation in visible light.

Below: the stages in the formation of our Galaxy.

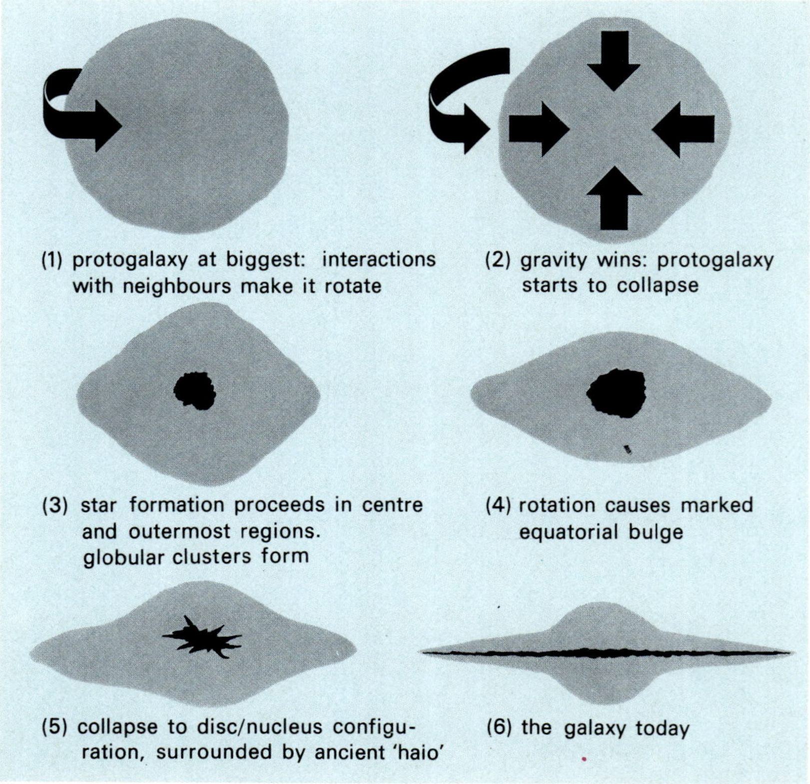

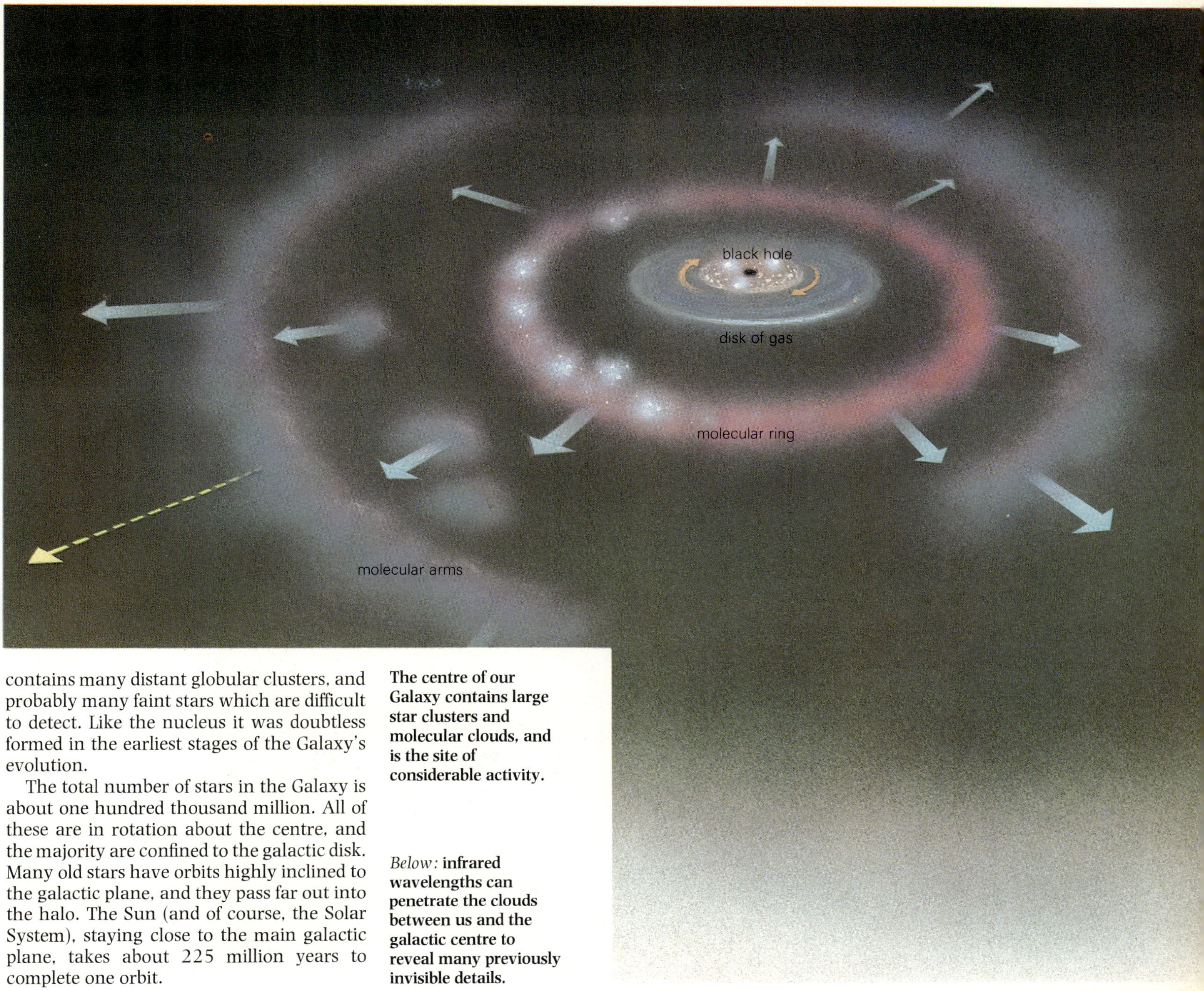

contains many distant globular clusters, and probably many faint stars which are difficult to detect. Like the nucleus it was doubtless formed in the earliest stages of the Galaxy's evolution.

The total number of stars in the Galaxy is about one hundred thousand million. All of these are in rotation about the centre, and the majority are confined to the galactic disk. Many old stars have orbits highly inclined to the galactic plane, and they pass far out into the halo. The Sun (and of course, the Solar System), staying close to the main galactic plane, takes about 225 million years to complete one orbit.

The centre of our Galaxy contains large star clusters and molecular clouds, and is the site of considerable activity.

Below: **infrared wavelengths can penetrate the clouds between us and the galactic centre to reveal many previously invisible details.**

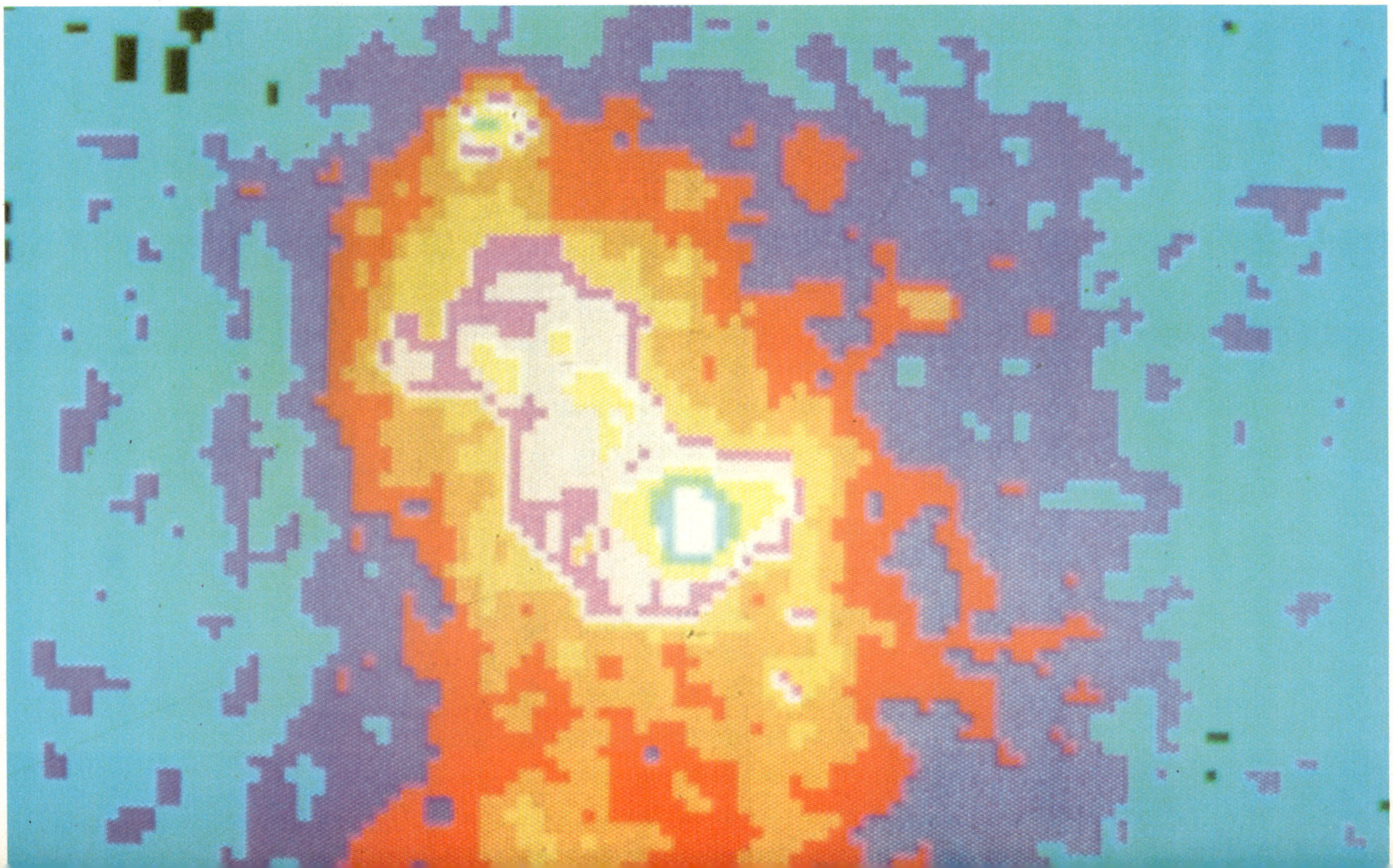

Galaxies

It was not until well into the 20th century that it was finally established that the 'spiral nebulae' were true 'island universes' like our own Galaxy, and that each consisted of many thousands of millions of stars. With the installation of large reflecting telescopes, particularly the 2·5m at Mount Wilson, it became possible to examine these galaxies in detail, to classify them, and to begin to determine their distribution in space.

Galaxies may be broadly divided into **spirals** and **ellipticals.** Spiral systems are like our own and also the nearest large galaxy, the Andromeda Galaxy (M31). They have a central nucleus of older, redder stars, and several spiral arms in which young blue stars, gas and dust are concentrated. In **barred spirals** the arms appear to arise from the ends of a more or less straight bar across the centre.

Elliptical galaxies – more properly called ellipsoidal galaxies – consist of just old stars with little gas and dust. They range in shape from spherical objects to highly flattened lenses, and in size (and mass) they vary far more than spirals. The giant ellipticals may be 100 times the mass of our Galaxy, while tiny dwarf ellipticals may have only one millionth of its mass.

Other galaxies can only be classed as **irregular**; the Small Magellanic Cloud is usually considered to be a fairly typical example, although it could have slight barred spiral structure.

Galaxies tend to occur in groups, or clusters. These may be small, as in the case of our Local Group, which is an irregularly-shaped collection of 20-odd galaxies, mostly dwarf systems, dominated by the two large spirals of the Andromeda Galaxy and our own. Like most large galaxies, both have smaller satellite systems.

Very much larger clusters of galaxies occur, such as the Virgo cluster, another irregular type, but with about 2500 members. Other clusters may be a more regular spherical shape, usually centred on a giant elliptical galaxy. At very great distances this

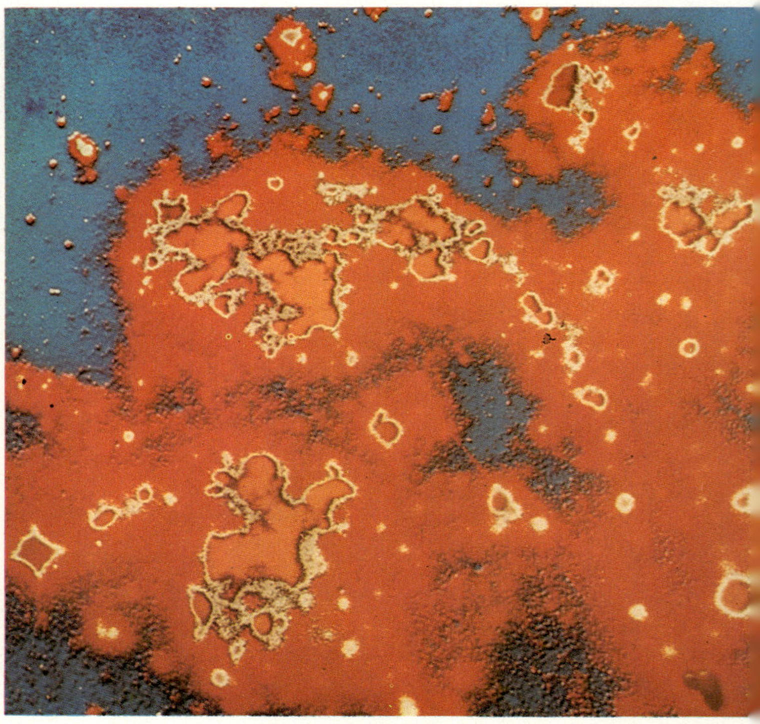

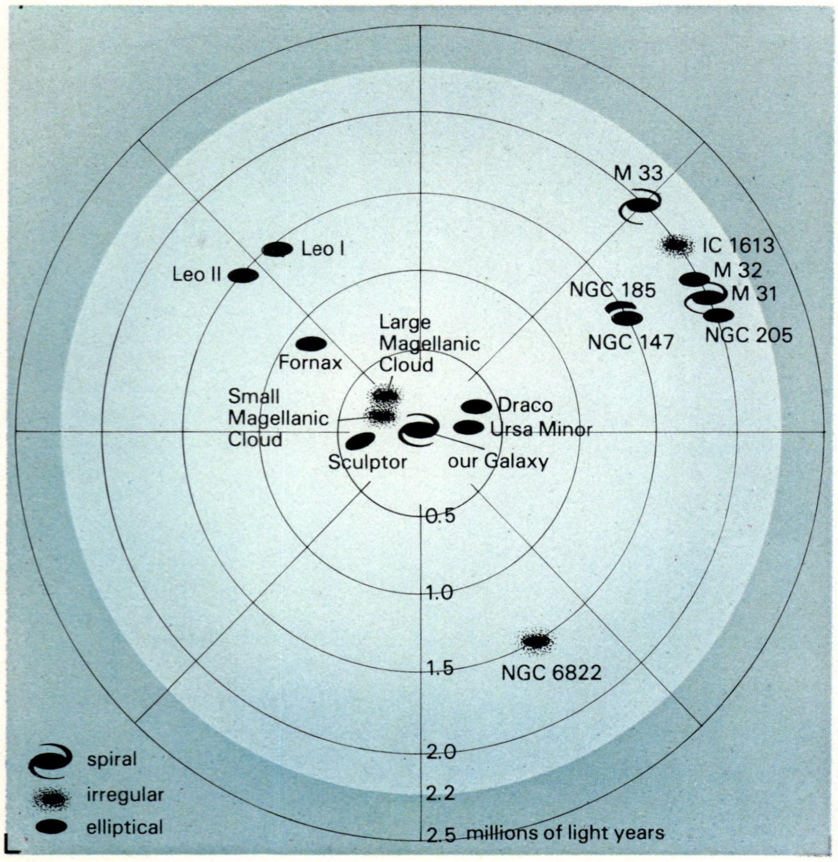

giant galaxy may be the only one readily visible.

Many galaxies are highly active and are associated with powerful radio sources. But the most violent activity is found in the **quasars**, which emit prodigious amounts of energy; and although they *appear* to be tiny, they are in reality far out in the universe. A few have now been shown to be associated with galaxies, usually difficult to see due to the extreme distances involved. It is thought that a quasar is actually the violent active central region of a galaxy, especially a young galaxy, more of which we see as we look further out into the universe, and further back in time.

Both ordinary and barred spiral galaxies may have either closely wound, or more open stellar arms.

Below right: **the Hubble classification of galaxies.**

These two galaxies, known as 'The Antennae', have been distorted by a near-collision.

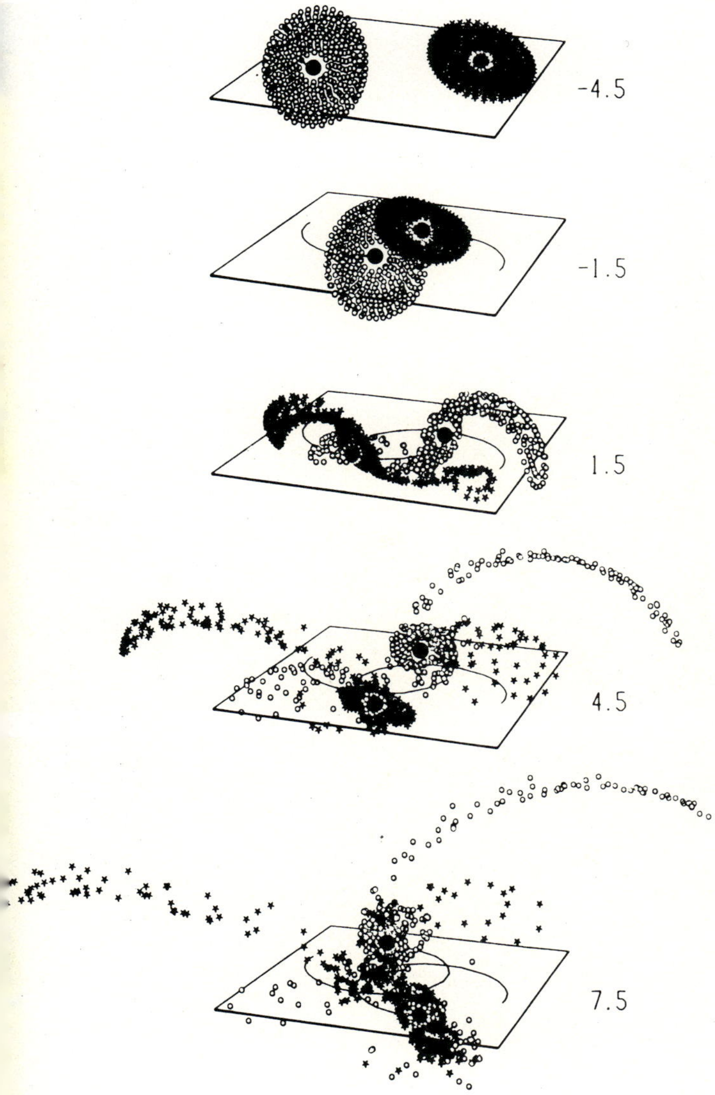

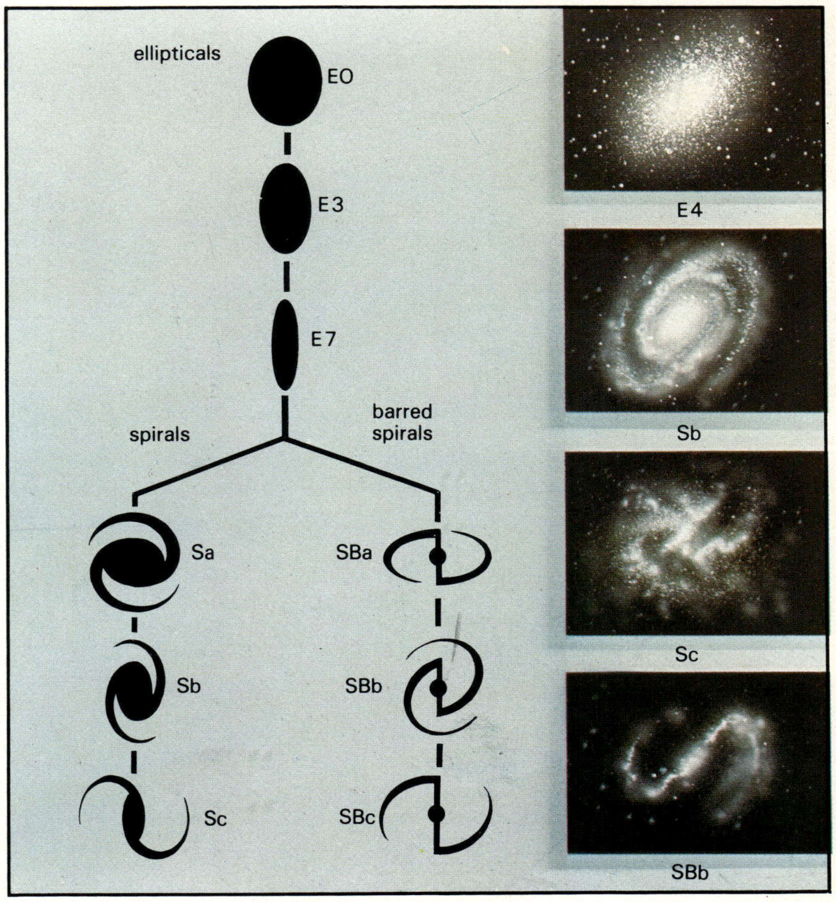

Opposite top left: **NGC 6997 is a striking spiral galaxy with young blue stars in the arms, and older red stars in the nucleus.**

Opposite top right: **the Large Magellanic Cloud, the nearest system, as it appears photographed in ultraviolet light.**

Far left: **the Local Group of galaxies is dominated by M31, the Andromeda Galaxy, and our own Milky Way system.**

Left: **M66 is a large spiral galaxy lying about 30 million light-years away.**

The expanding universe

One of the most useful pieces of information which can be gained from an examination of the spectra of astronomical objects relates to their motion. If a star or other body moves towards an observer, the dark or bright lines (see page 48) will be displaced from their normal positions and shifted towards the blue end of the spectrum. Conversely, red-shift indicates motion away from the observer, and in both cases the amount of the displacement indicates the velocity. The motion of individual stars and the rotation of galaxies are just two of the many facts which can be detected from this **Doppler effect** (named after the discoverer).

When the spectra of galaxies were examined it was found that with a few minor and very local exceptions, all galaxies showed a

Below: **viewed from 'outside' some galaxies would show blueshifts, but from within our expanding universe only redshifts are encountered.**

Below: **the greater the distance of the galaxy, the greater its apparent velocity of recession.**

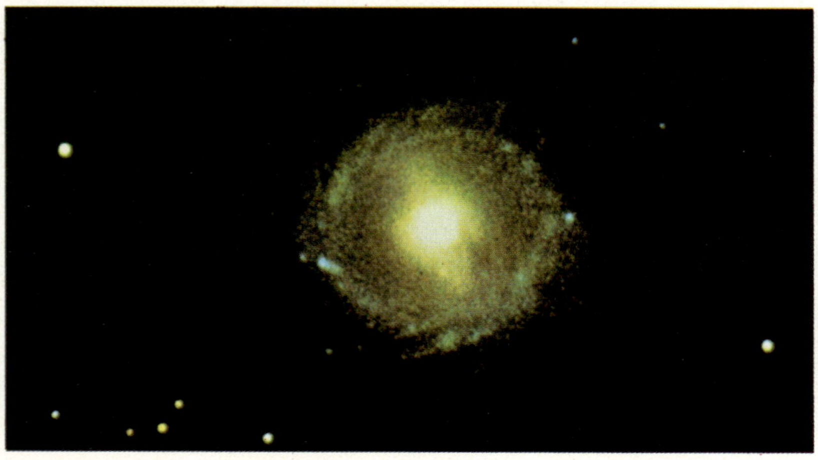

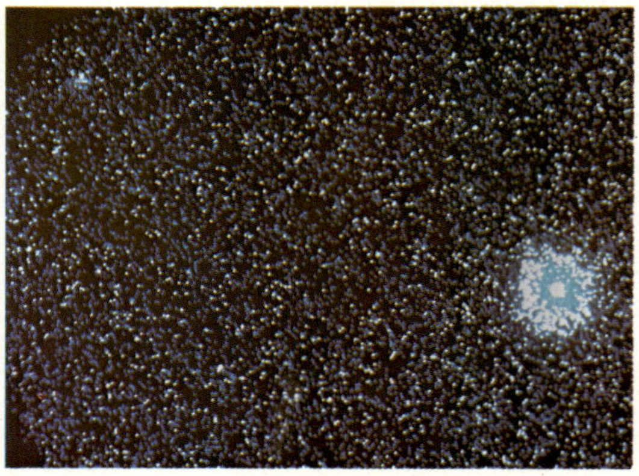

redshift, and moreover that this became greater with increasing distance, implying that the velocities increased similarly. The relationship still holds for the far more distant quasars, some of which have recession velocities which are a very substantial fraction of the speed of light. The farthest yet detected quasar (known as PKS 2000–300), exceeds 90 per cent of the speed of light, and must lie at a distance of about 13 000 million light-years.

At first it might appear that the increase in velocity with distance implies that our Galaxy is at the centre of the Universe from which all other galaxies and quasars are receding. However, this is not the case, and the appearance from any one of the other systems would be just the same. It is the entire universe which is expanding.

When and how did this expansion begin, or is it just an illusion? Various cosmological theories—theories of the Universe as a whole – suggest that it did have a true beginning when the entire Universe consisted of one single, unimaginably dense, 'superatom' which then exploded. This **Big Bang**, the origin of the Universe, was immensely hot, and although now cooled by its vast expansion, has left a trace of that high-temperature stage in the background radiation detectable by sensitive radio equipment. This tiny trickle of heat comes evenly from all directions, a condition which is so difficult to explain by any other theory that it provides strong confirmation that ideas of a Big Bang are correct.

Will the Universe expand forever, or will it slow down, stop, and then collapse again? Does it oscillate from one Big Bang to another, and so on for ever? Astronomers with their ever more complicated equipment, and cosmologists with their ever more complex theories, struggle to provide the answers. For the present though, all we can say is that the origin of our Universe lies in that Big Bang, fifteen thousand million years ago.

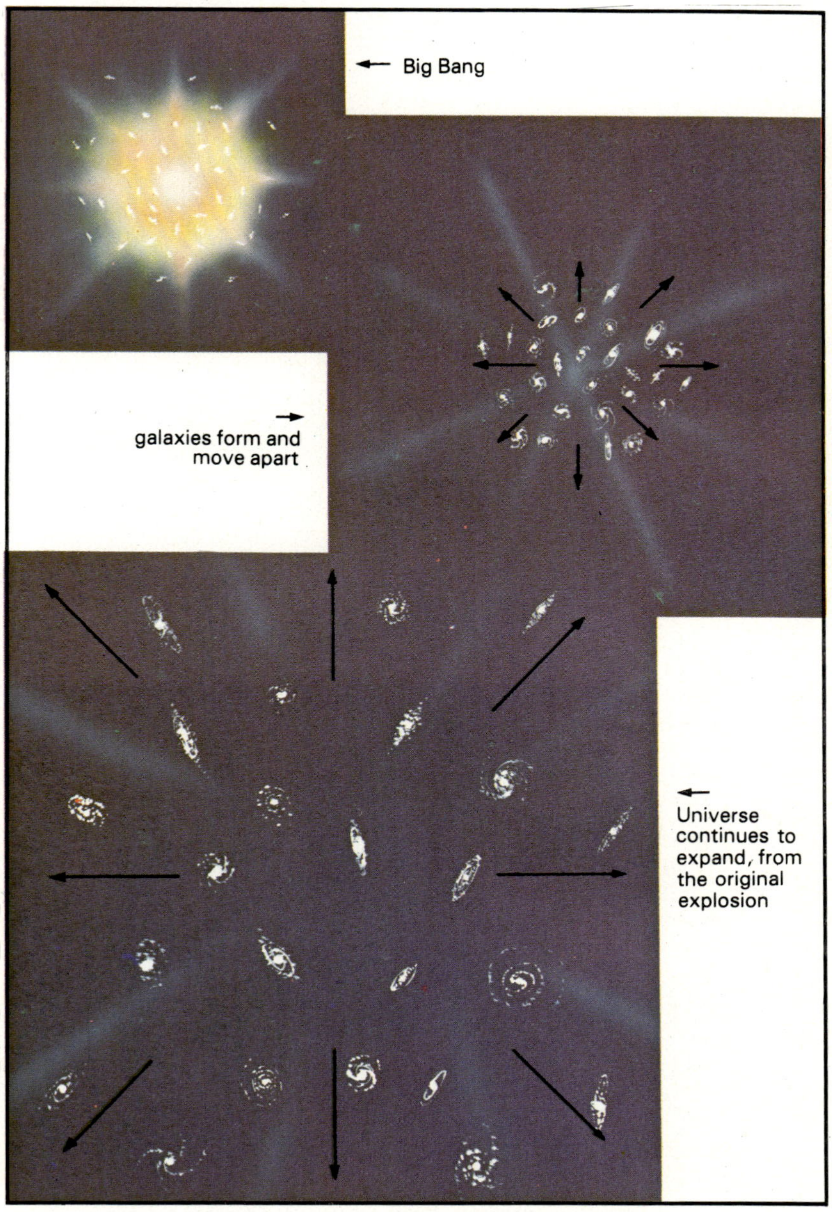

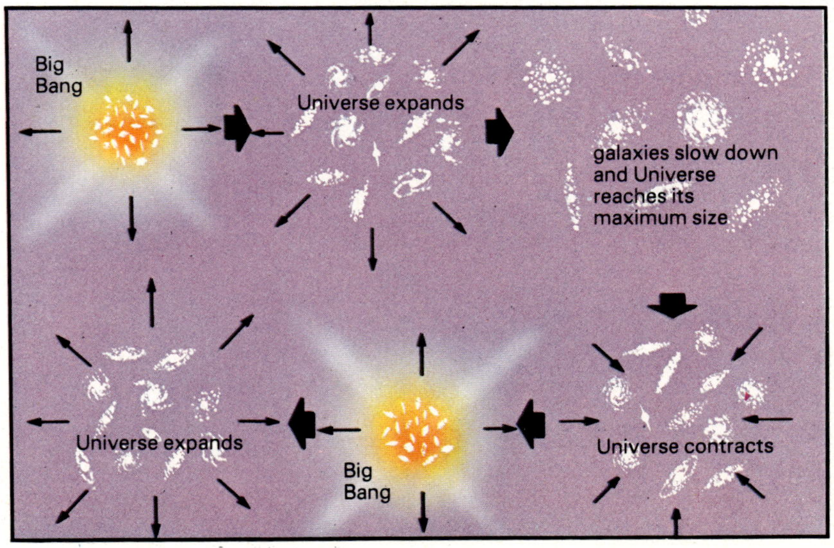

Opposite bottom left: **M95, a galaxy in Leo, has a redshift corresponding to a velocity of about 600km per second.**

Opposite bottom right: **the quasar 3C-273 observed in X-rays by the Einstein satellite.**

Top right: **quasars are thought to be the violently active centres of very young galaxies.**

Centre right: **the whole universe probably began in a single explosion, the 'Big Bang', from which it is still expanding.**

Right: **in the 'Oscillating Universe' theory, the Universe may go through many cycles of expansion and contraction.**

Telescopes and observatories

Although astronomers nowadays use some very strange and sophisticated equipment in their study of the universe, the main tool remains the large-diameter telescope. Most of these are reflectors, the successors of that crude original built by Sir Isaac Newton. Although still using mirrors for the primary purpose of gathering as much light as possible and bringing it to a focus at a convenient position for examination, a large number of specialized designs have been developed for specific tasks. These may require various shapes of mirrors to achieve the necessary performance, or else need mountings specially designed for the task in hand. For example, a telescope required for the determination of stellar positions – and perhaps for the determination of distances, a notoriously difficult task – must be very rigid indeed, so that the telescope's flexure does not introduce errors greater than the tiny amounts which are to be measured.

Most large telescopes will only cover a very restricted field of view. This may be quite satisfactory for the study of individual galaxies, but is completely useless for studies of the distribution of galaxies in space, so essential to the investigation of various cosmological theories. For such wide-field coverage, the Schmidt telescope design is used, and this may record thousands of galaxies on a single photographic plate.

In order to follow the movement of astronomical objects, telescopes must be mounted in such a way that the rotation of the Earth may be counteracted. For many decades this has meant that all large professional telescopes have used a large, complex equatorial mounting, where one of the two main axes is parallel to the Earth's rotational axis, to give this compensation by a single, simple drive. Any other form would require rotation on two axes, and probably rotation of the auxiliary equipment as well. But now computers can handle this with ease, and all new telescopes will probably appear with the 'old-fashioned' altazimuth design. The telescopes and enclosing domes may be lighter, smaller – and cheaper!

Astronomers have always had a problem finding the most favourable sites for their telescopes and observatories. Generally it is the atmospheric conditions which largely determine how frequently a telescope may be used, as well as the quality of the seeing during the night-time hours. It is for this reason that most modern observatories are sited at the tops of mountains, high above the usual cloud levels. With the development of satellite communication links, and sophisticated remote control, it is probable that many new telescopes will be operated by astronomers at home in their own countries, not from the observatories themselves.

The Anglo-Australian Telescope (3·9m in diameter) is typical of modern large reflectors.

Below: **the sites of major observatories and large optical and infrared telescopes.**

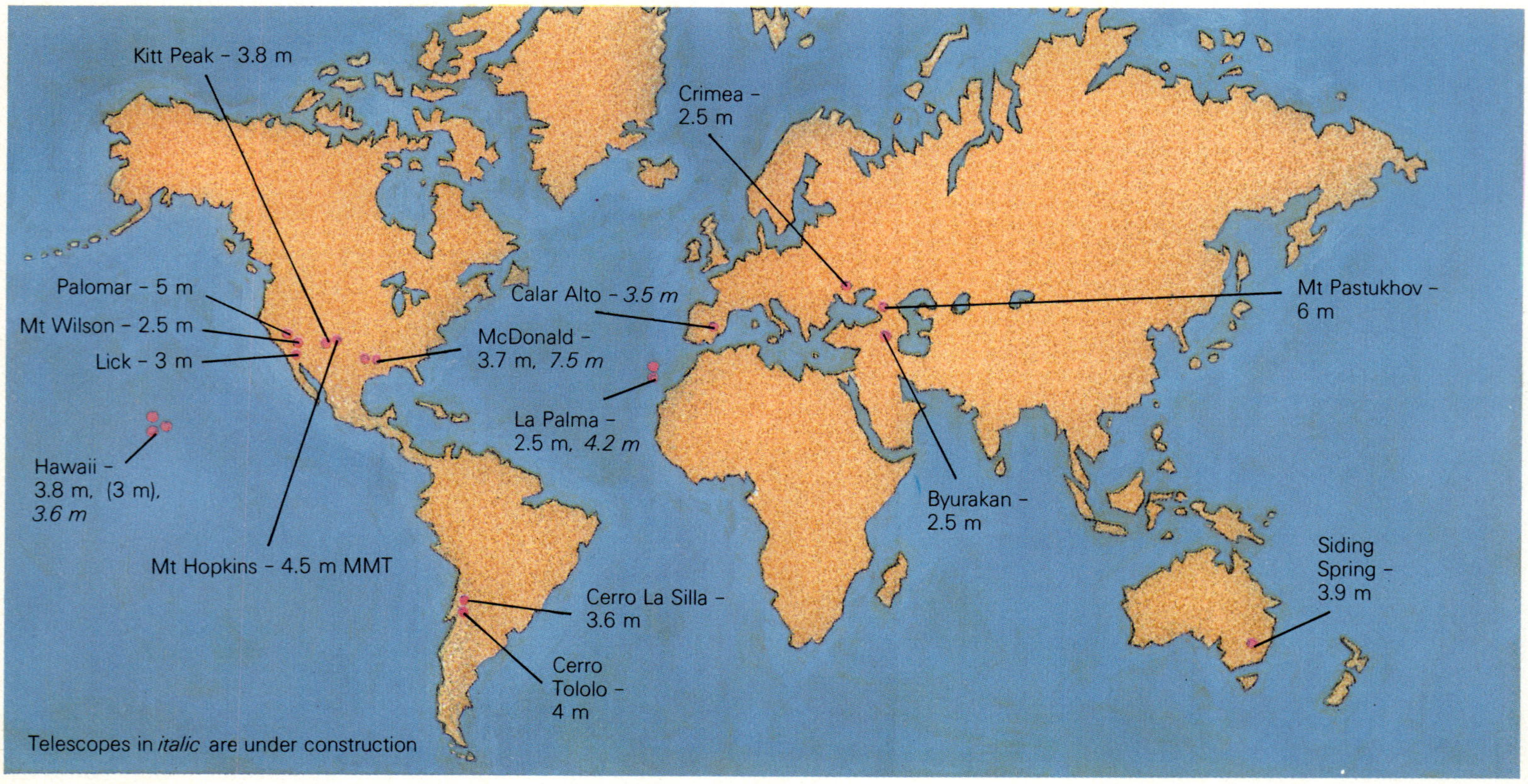

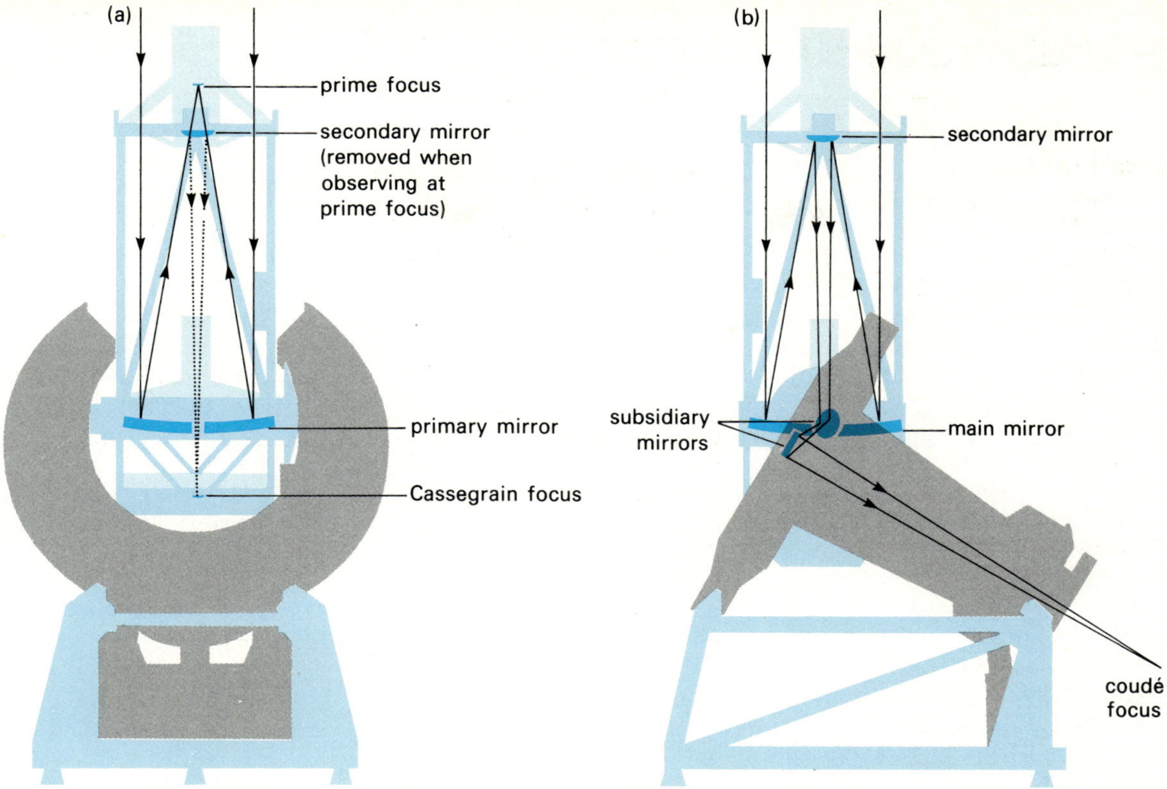

The complex equatorial mounting required for large optical telescopes.

Right: the Multi-Mirror Telescope heralds a new form of telescope using many individual mirrors.

Below: the 91-cm telescope at Lick Observatory is the second-largest refractor in the world.

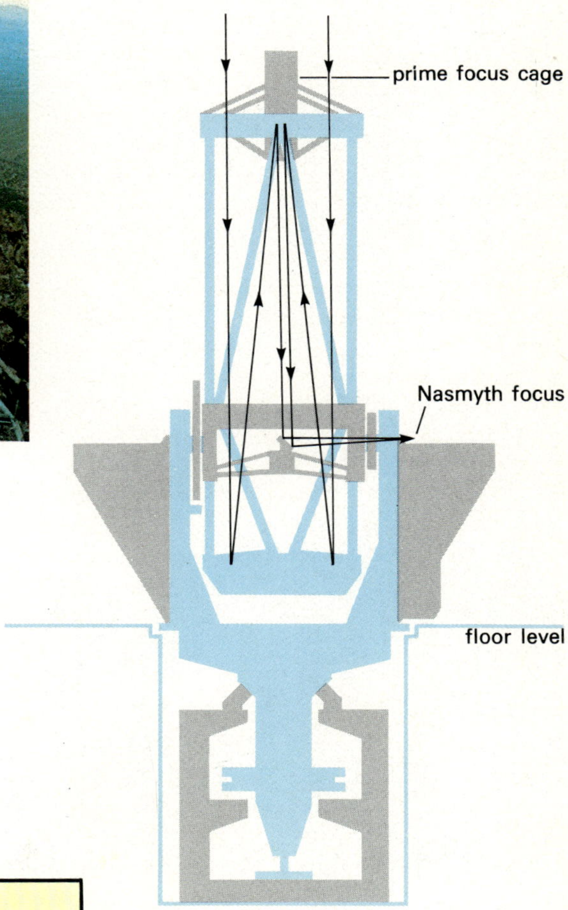

Even the largest modern telescopes (this is the Soviet 6-m) may have alt-azimuth mountings.

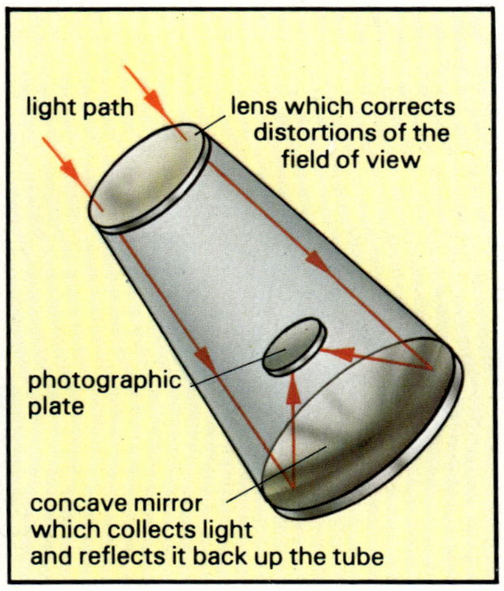

Left: Schmidt telescopes are designed to photograph very wide areas of sky in a single exposure.

Invisible astronomy

Astronomical observations now extend far beyond the visible region of the spectrum, and it is difficult to realize why the first observations of radio waves from celestial objects caused considerable surprise. **Radio telescopes** are now quite familiar to everyone, even if only because they have sometimes been used to communicate with distant spacecraft. Complex techniques are used to interconnect radio telescopes across continental distances, and achieve (in some cases), finer resolution than optical telescopes, revealing both tiny details in the centre of our Galaxy and gigantic radio lobes of far distant quasars.

Infra red observations too, have opened up new possibilities for the study of many objects, from planets and new-born stars to distant galaxies. But the Earth's atmosphere blocks much of the infrared radiation, so high-altitude observations such as Mauna Kea in Hawaii (altitude 4267m), or satellite-borne equipment are essential.

On the shorter-wavelength side of the optical 'window', **ultraviolet** observations carried out from space bring information about the brilliant, most massive stars, as well as providing insight into many other energetic systems. Similar information comes from **X-ray** observations of objects from the most distant quasars to the Sun itself. Finally, even **gamma-rays,** the most energetic radiation of all, are beginning to yield details of some known astronomical objects, as well as recording the enigmatic 'gamma-ray bursters' – strange pulses of gamma-rays of unknown origin.

Astronomers are concerned with other observations as well. There are, for example, the **cosmic rays** – highly energetic particles which smash into the top of the Earth's atmosphere and cause showers of secondary particles to be produced. Some originate as close as Jupiter's magnetosphere, but others have been produced by supernovae and accelerated by the Galaxy's magnetic field.

Stranger yet, astronomers are at work deep underground. Theory predicts that the nuclear reactions powering the Sun and other stars should produce floods of ghostly particles called **neutrinos,** which normally pass straight through the Earth. To stand any chance of detecting them, equipment must be shielded from all other radiation and particles by thick layers of rock. Current results indicate that the theories of the Sun's interior need revision.

Finally, there is the search for **gravitational waves.** In theory these are produced by any accelerated mass, but in practice they would be so weak that in most cases they would be completely undetectable. Only in the case of exotic objects and violent events, such as the collapse of a star into a neutron star or black hole, is there any hope of their detection. Several teams are working to see if any evidence for the existence of gravitational waves can be found, but so far with no success.

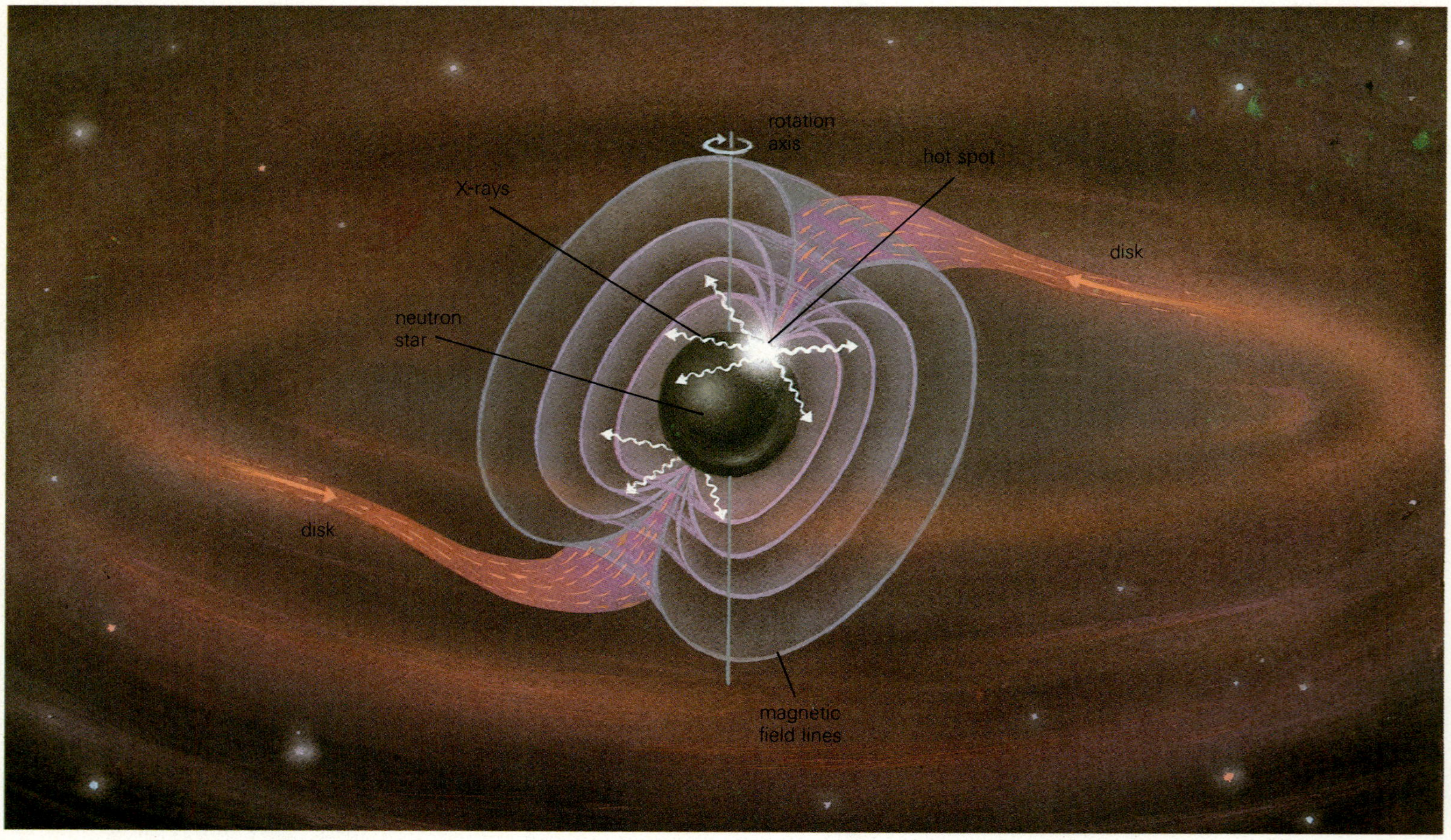

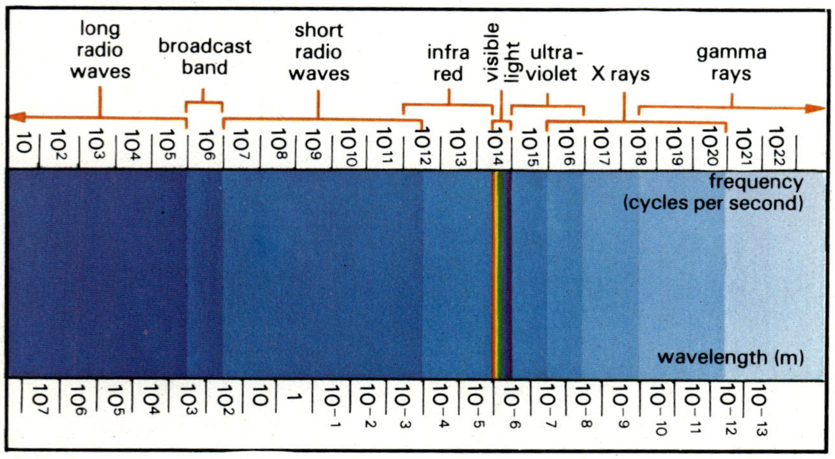

Above: **matter flowing down onto a collapsed object gives rise to the radiation from X-ray stars.**

Left: **recent advances mean that astronomers may now observe over the whole of the electromagnetic spectrum.**

Opposite top left: **this is a view of the giant galaxy M82 as seen in X-rays.**

Opposite top right: **the large radio telescope at Parkes in Australia is used for many studies of distant galaxies.**

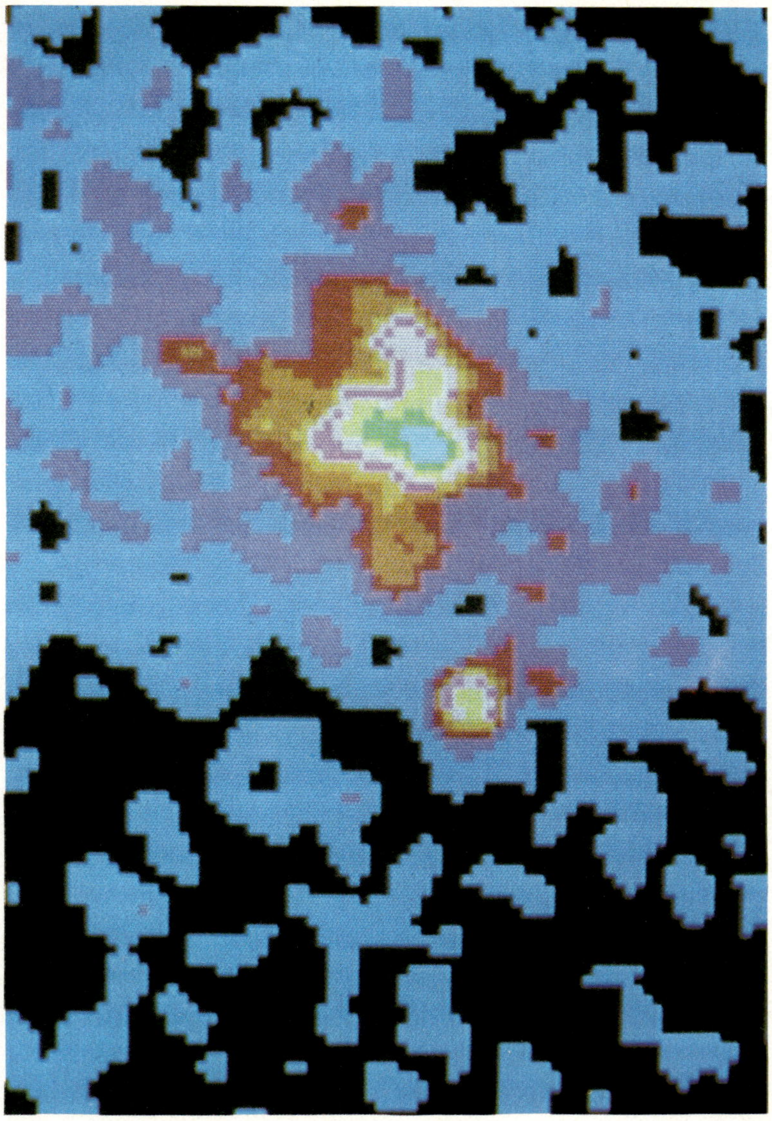

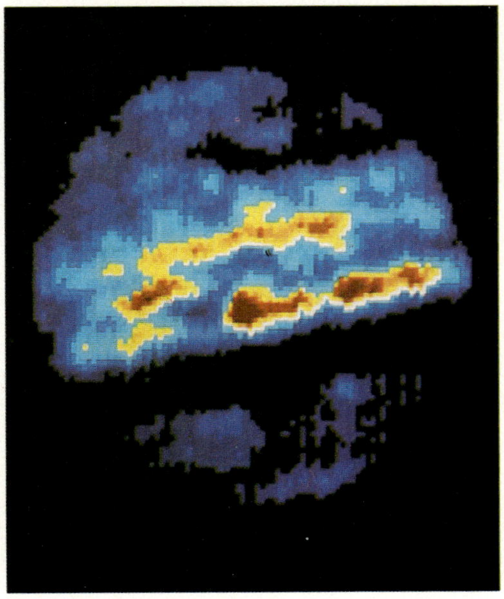

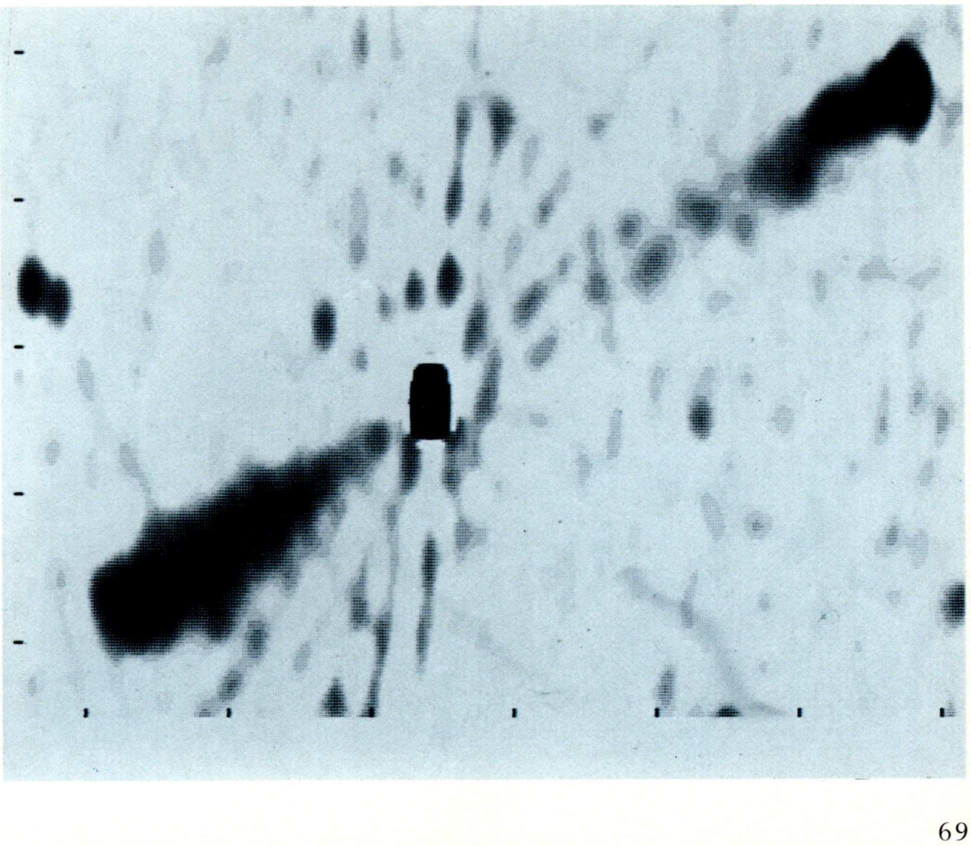

Above: an infrared image of the planet Jupiter, showing that the belts are the hottest regions.

Above centre: HEAO-1 (High Energy Astronomical Observatory) was the first major X-ray satellite.

Above right: the highly successful IRAS (Infrared Astronomical Satellite) observatory, launched in 1983.

Right: the giant radio lobes of galaxy 3C-236 are more than 15 million light-years across.

Astronomy from space

The giant Saturn V rocket was needed to take astronauts to the Moon, but experiments in Earth orbit will be launched by vehicles like the Space Shuttle and Ariane.

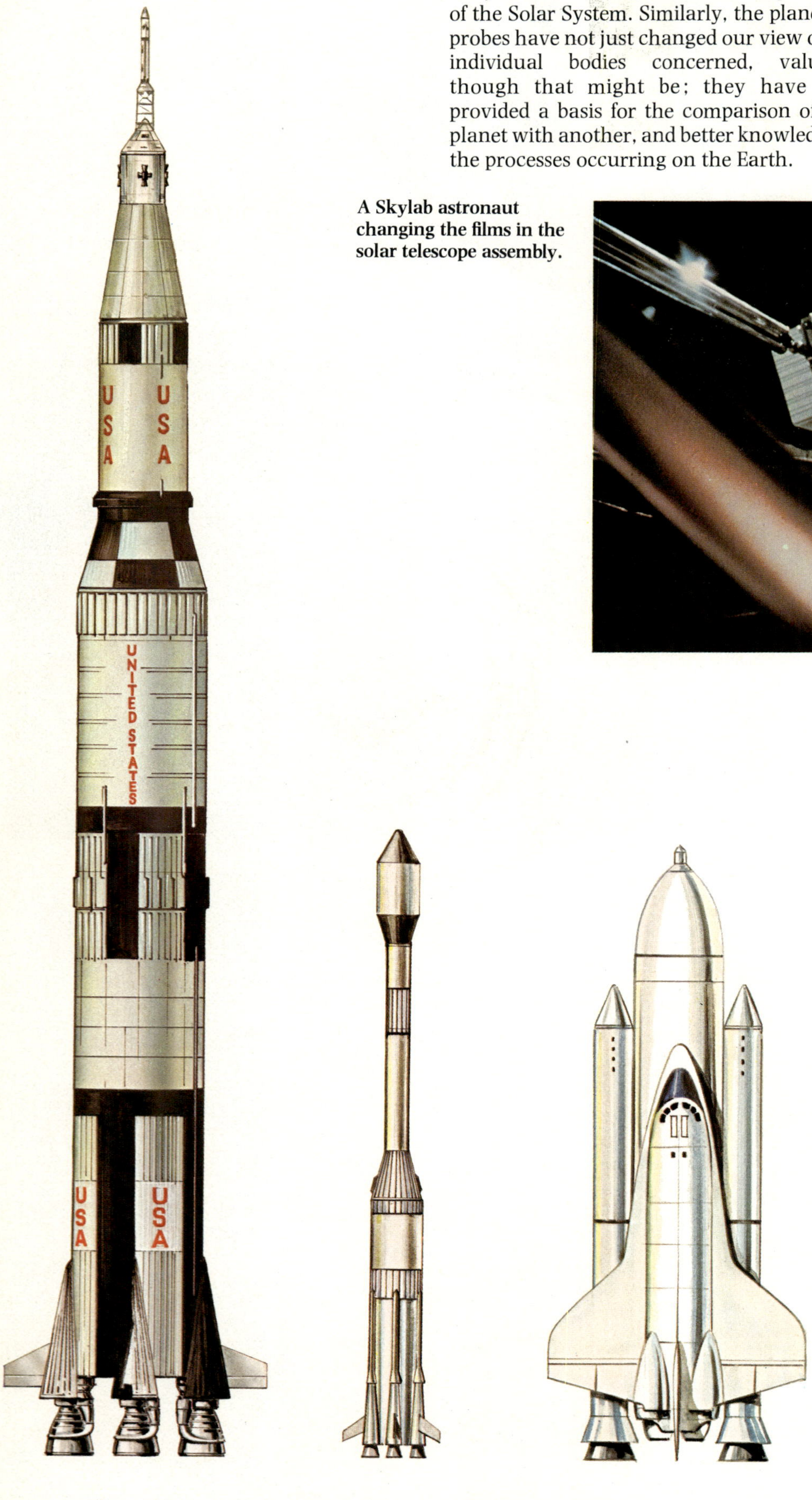

Space research has had an almost unbelievable impact on astronomy since it first began in 1957. For astronomers, the Apollo programme was remarkable not for landing men upon the Moon, but rather for – among other things – returning rock samples which have completely changed our understanding not just of the Moon, but of the early history of the Solar System. Similarly, the planetary probes have not just changed our view of the individual bodies concerned, valuable though that might be; they have also provided a basis for the comparison of one planet with another, and better knowledge of the processes occurring on the Earth.

Clear of the interference of the Earth's atmosphere, objects can be studied at the wavelengths which are most suitable, and in such a way as to give a complete picture of their radiation. The Sun, for example, may be studied continuously from satellites in Earth orbit (as was done from Skylab), from long-lived monitoring by solar-orbit probes, or from short-term experiments, perhaps carried aloft by the Space Shuttle. Our understanding of its effects upon the Earth has already changed greatly, and information about its long-term effects upon our climate will begin to become available.

A Skylab astronaut changing the films in the solar telescope assembly.

Not only can known bodies be investigated freely from space, but strange, previously unknown objects have also been discovered. We may begin to understand X-ray stars, and the signs that a black hole may show; but other phenomena such as gamma-ray bursters, for example, remain a mystery.

What may we expect in future? That is difficult to foresee, but many projects now in planning should bring yet more advances. The Space Telescope, for example, may be able to detect planets of other stars, as well as offering clear views of the most distant objects in the Universe. The first infrared survey of the sky has been completed by IRAS (Infrared Astronomical Satellite), and later craft should consolidate that knowledge. The Hipparchos satellite will provide stellar positions of unprecedented accuracy, and forthcoming X-ray satellites should enable some of the enigmatic sources to be examined in detail. We have yet to see radio telescopes in orbit, but they are projected, and could well be linked to those on Earth to make simultaneous high-resolution observations.

Space probes will also bring us new knowledge of cloud-enshrouded Venus and Jupiter, of Uranus perhaps, of Comet Halley, and of the Sun itself. Missions to minor planets, mapping of the hidden surface of Titan, sample returns from Mars, and many more scientific dreams will remain for the future to fulfil.

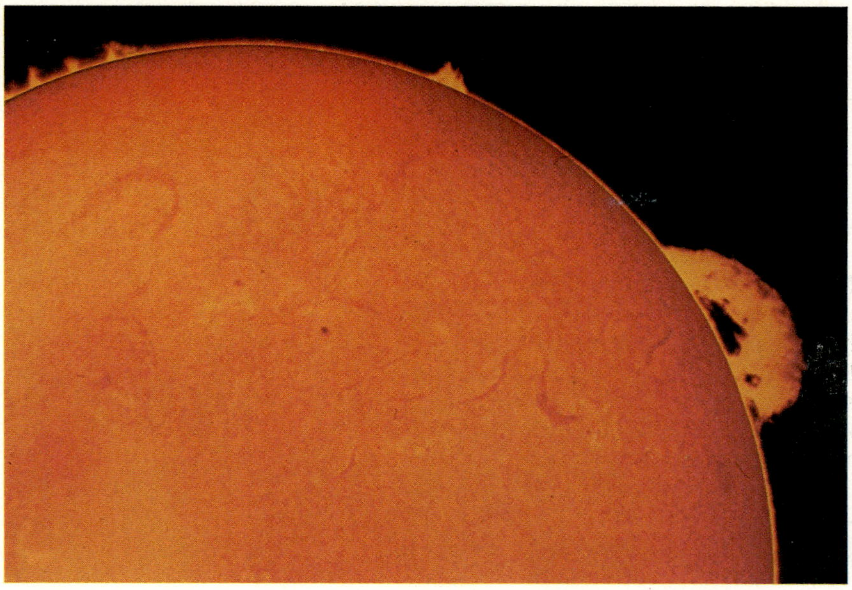

It is only from space that full details of the Sun can be observed.

Far right: the Voyager spacecraft which studied Jupiter and Saturn with such success.

Below: Mariner 9 first returned detailed maps of the surface of Mars.

Right: the Space Telescope should bring greatly improved knowledge of many astronomical objects.

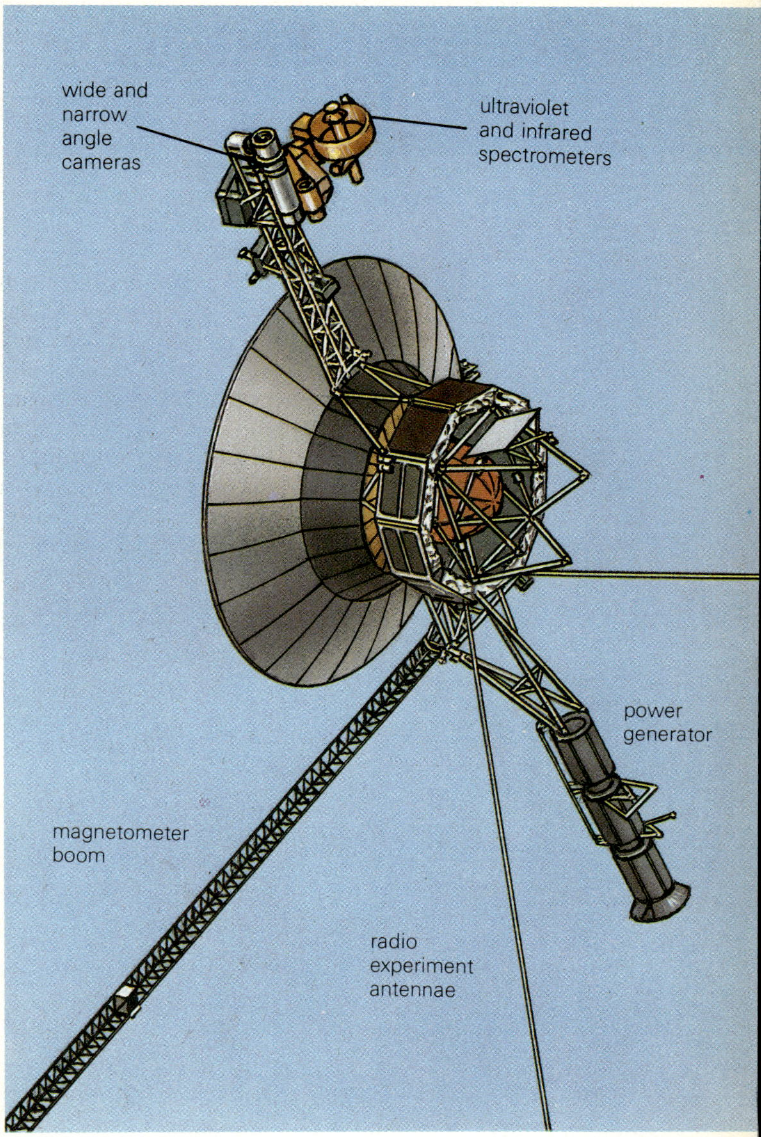

Amateur observing

There have always been many amateurs interested in astronomy, from the casual 'ramblers' to the highly dedicated 'scientific' observers. Perhaps surprisingly, some of the latter may not use any equipment at all for some of their work, as in the observation of meteors, for example. Casual observers will find much that is rewarding to do, from watching a lunar eclipse to learning their way about the sky.

Obviously some equipment is often useful, even if not essential. Any amateur astronomer can make good use of a pair of binoculars, and these are probably a better investment than most of the small (50mm or thereabouts) refractors, which are frequently offered for sale at tempting prices. Moreover, binoculars can not only be used for other objects during the day, but remain very useful even after one has acquired a larger telescope.

What should that larger telescope be? For preference a reflector of about 150mm diameter; a 75mm refractor (which would

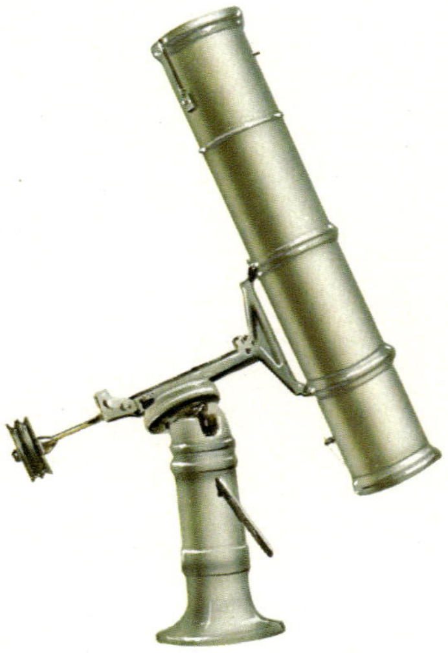

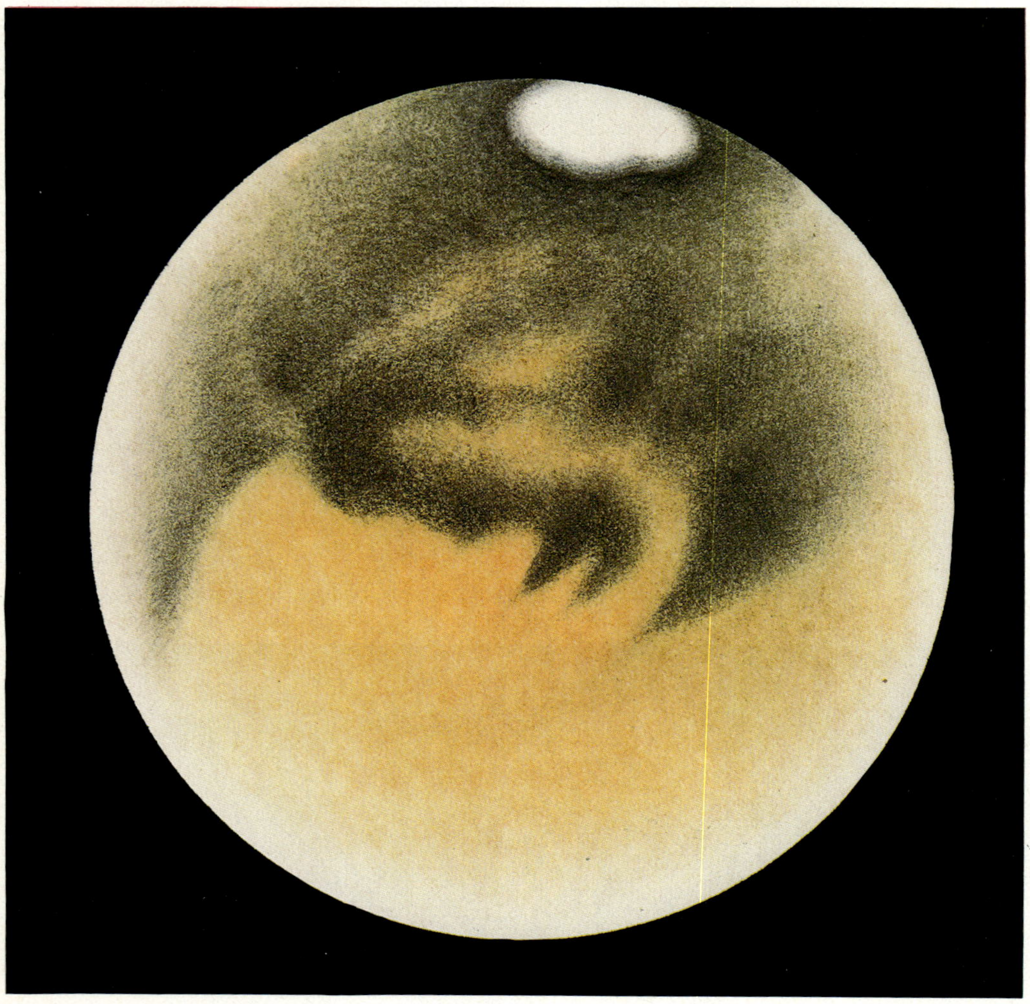

Left: **a detailed observation of Mars, showing many details, including a polar ice cap.**

Above: **the Newtonian reflector is one of the most popular types of telescope with amateur observers.**

Below left: **the only safe way of observing the Sun is by projection onto a white screen.**

Below right: **The 7 x 50 size of binoculars is ideal for many amateur observing programmes.**

Opposite top right: **an equatorial mounting allows easy compensation of the rotation of the Earth.**

Opposite centre: **the lunar crater Aristarchus and** *(inset)* **the Straight Wall.**

Opposite bottom left: **this light-curve of the long-period variable Mira has been drawn from amateur observations.**

Opposite bottom right: **the Newtonian** *(left)* **and Cassegrain forms are those most frequently used by amateurs.**

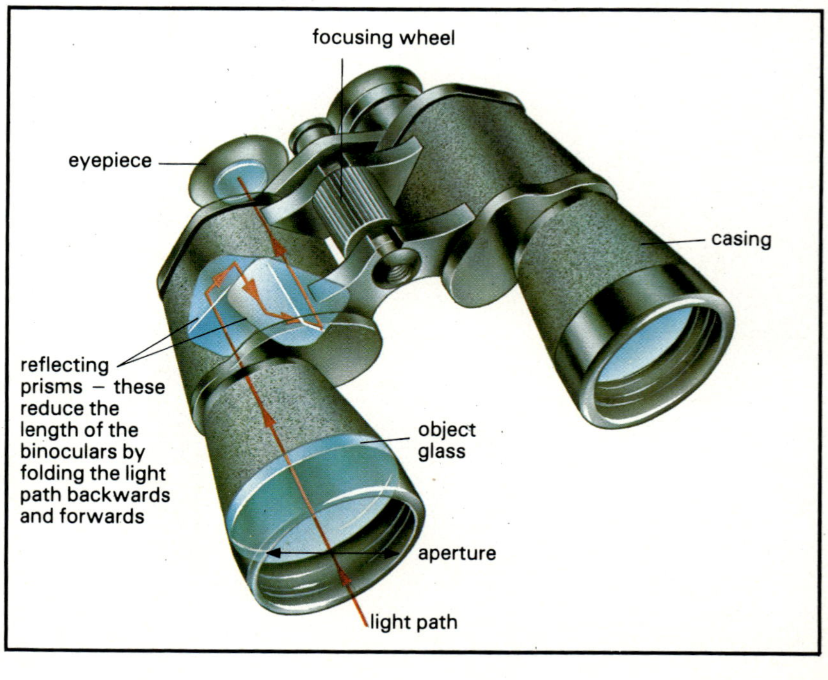

Left: **the Moon, Venus and Jupiter** *(left)*. **A one-second exposure on 160 ASA film, with a 50mm, f/1·4 lens.**

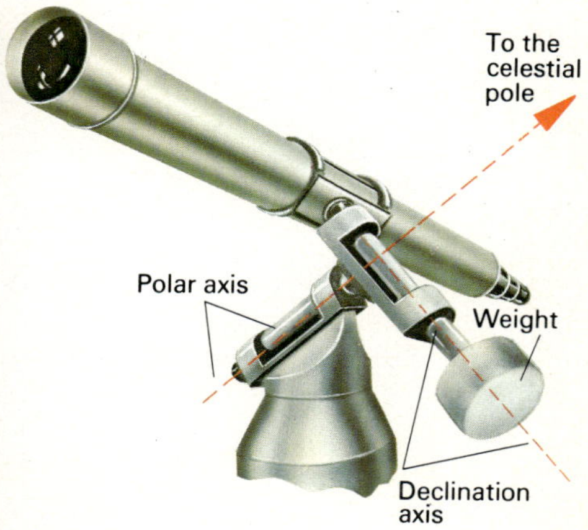

cost about the same) is about the smallest telescope that could be recommended. Either of these could afford a lifetime of pleasurable study, although probably the wish to make *useful* observations would soon become apparent. Many amateurs derive great pleasure from making their own telescopes (usually reflectors), and although in most cases they purchase the optical parts, they do obtain a better instrument for a reasonable price. Large commercially-built telescopes are expensive, so most are home-made.

Ordinary cameras may also be used to good effect and there are many designs for simple driven mountings, for both cameras and telescopes, which can be easily constructed.

What do the 'serious' amateurs observe? Practically everything – that is the only answer. In some fields they make a major contribution to science: observations of aurorae, meteors, variable stars and the markings of Jupiter are just a few examples. They make searches for (and discover) comets, novae and supernovae, all of which are of the very greatest interest to professional astronomers. But none of this is intended to belittle those who may just be interested in seeing if they can identify a crater on the Moon, split a double star, find Mercury in the twilight, or trace the clouds of the Milky Way.

Finally it is worth remembering that as the Sun is an astronomical object, some observations are possible in daylight; but it must not be forgotten that it must never be viewed through any optical instrument, however small. The only safe method is that of projection, which is quite simple, allows all the solar detail to be seen, and yet is completely without danger. Even the heat of a low Sun on the horizon can cause damage or blindness, so do take heed of this warning!

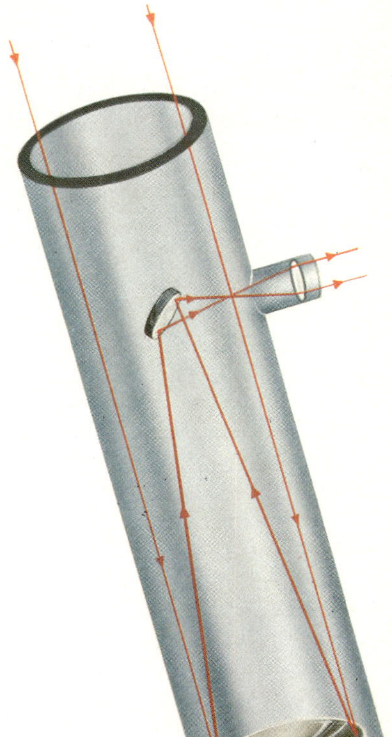

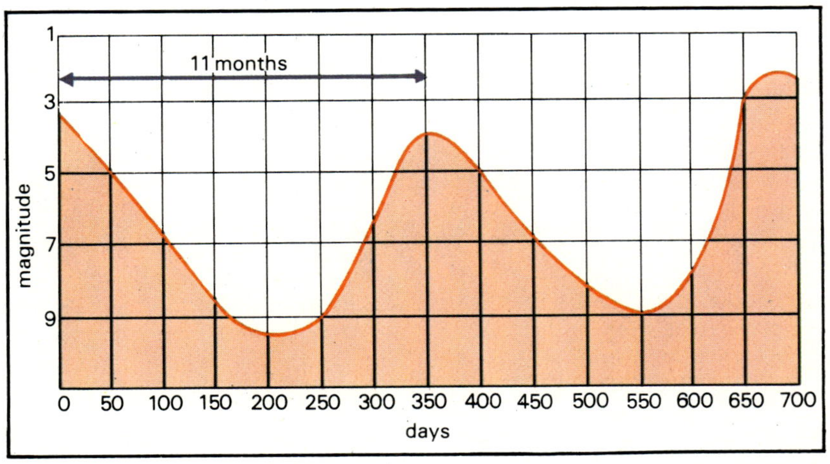

The constellations

The three stars of the 'belt' of the mythical hunter Orion, together with the 'sword' – the Orion Nebula.

Above right: **a map of the northern constellations as drawn by the 17th-century astronomer Johannes Hevelius.**

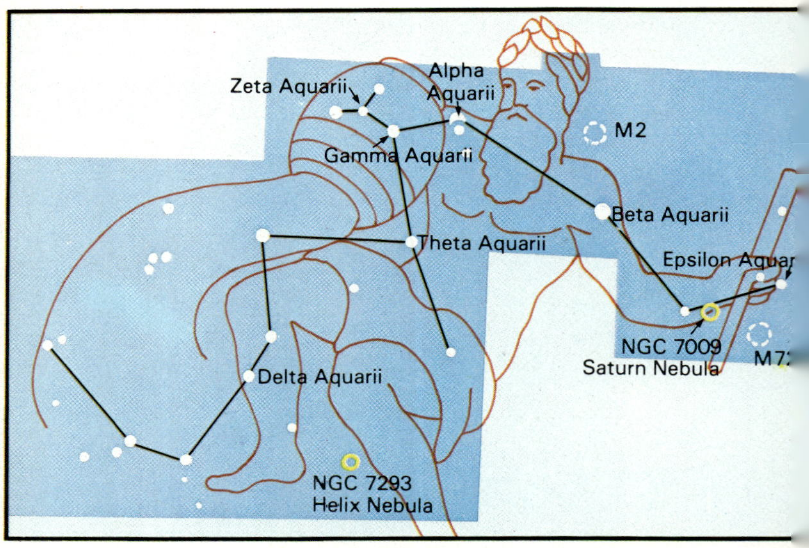

Above: **in ancient times Aquarius was envisaged as a man pouring water from a jar.**

There are so many stars in the sky that many beginners despair of ever being able to find their way around. But it has always been natural for groups of stars to be linked into constellations, and these will form a guide. Peoples around the world have devised many constellations, some consisting of such natural groupings that they have continued in use to the present time. The names of many were, of course, largely influenced by superstitious beliefs and were taken from those of mythological gods and legendary creatures, although some were of everyday objects and creatures. The names which we use today were only finally agreed in 1930. Most astronomers find it best to use the Latin names, rather than any translated versions.

Although at the outset – and indeed for many centuries – undertaken for astrological rather than astronomical purposes, the charting of the stars eventually became essential. In the West, with its concentration upon the Sun, Moon and planets, the system of co-ordinates usually employed was based upon the **ecliptic** (the Sun's apparent path in the sky). The Chinese, on the other hand, early adopted the system of **polar co-ordinates,** like the latitude and longitude used in geography, and used in the sky today. However, one fundamental reference point is still provided by the position at which the ecliptic crosses the celestial equator.

Many of the brighter stars had been given names in ancient times, again usually mythologically inspired. Over the years these were considerably modified, but many of those found today were given by the Arab astronomers of the Middle Ages. With increasing scientific awareness and the need for greater precision in navigation and timekeeping, it became obvious that many more stars needed to be adequately identified and their positions accurately noted. The German astronomer Johann Bayer introduced in 1603 the use of Greek letters (and some Roman ones) for the identification of the brightest stars in each constellation, and John Flamsteed, the English Astronomer Royal, in 1725, used numbers for many fainter stars. Both of these identification systems are still in use, as are Charles Messier's catalogue numbers (1784) for certain nebulae. Numbers from J. L. E. Dreyer's NGC (*New General Catalogue of Nebulae and Clusters of Stars*), based upon work by the Herschels, are frequently encountered, as are the later IC (*Index Catalogue*) revisions. Many later specialized catalogues are, of course, used among astronomers.

Originally the constellations had no fixed boundaries – these tended to depend on the whims of the person preparing the charts. Irregular, but generally accepted, boundaries were in use for a while, but at the same time as the constellation names were finally agreed, rectangular limits were defined and accepted.

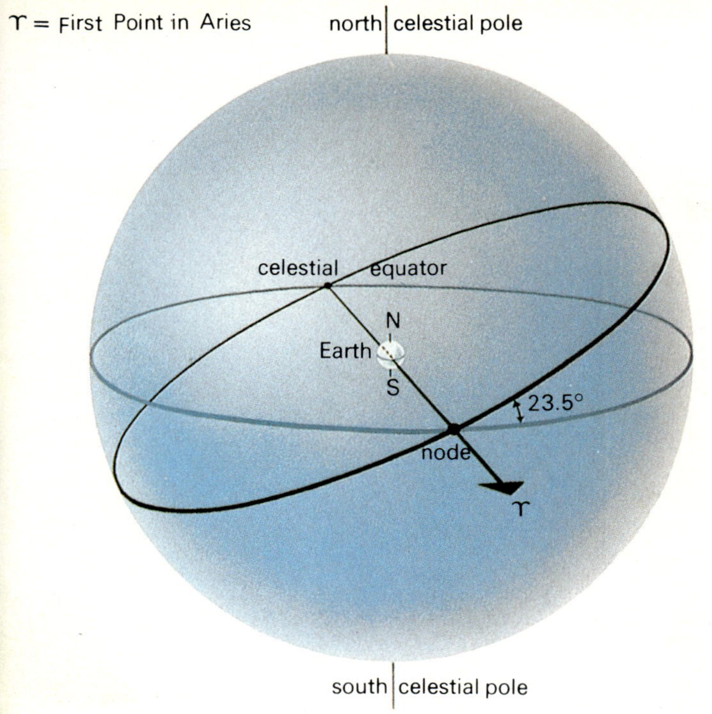

Top left: the First Point of Aries (γ) is where the Sun crosses the celestial equator from south to north.

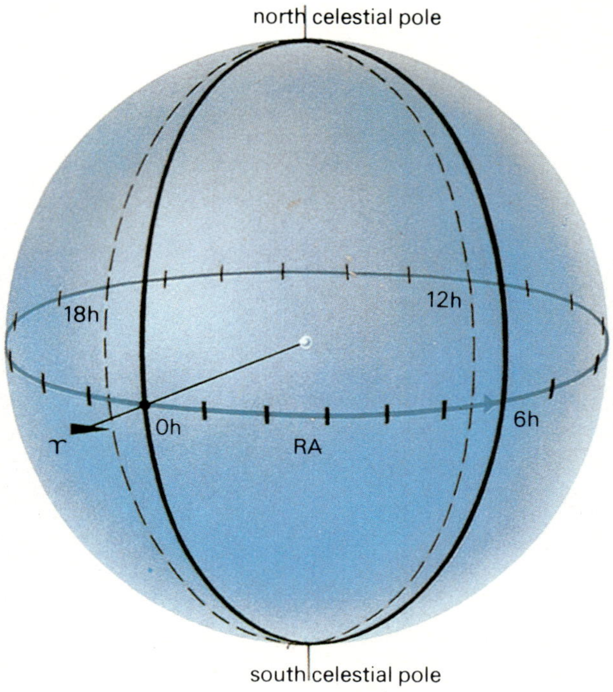

Left: Right Ascension is measured eastwards around the celestial equator from the First Point of Aries.

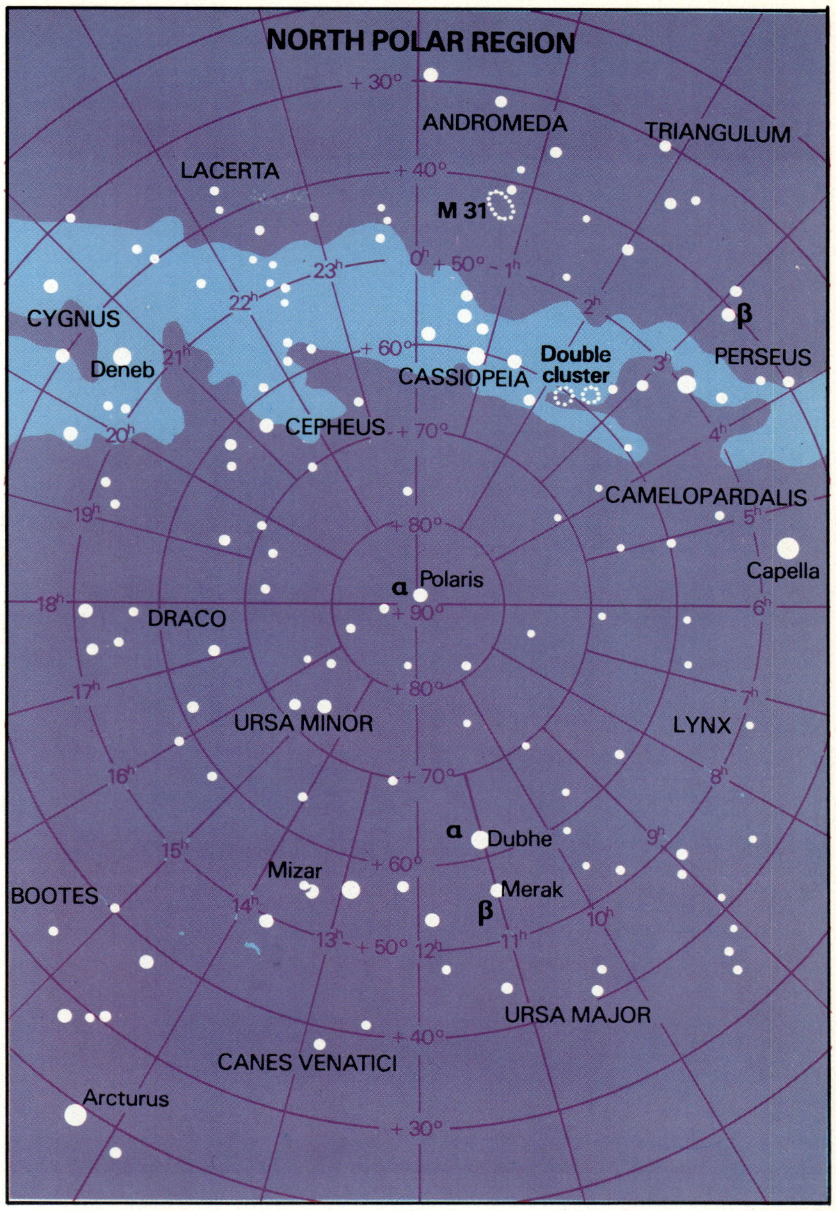

Above: the two Pointers in Ursa Major indicate the position of Polaris. The positions of the Double Cluster in Perseus and of the Andromeda Galaxy, M31, are also marked on this chart of the northern polar region.

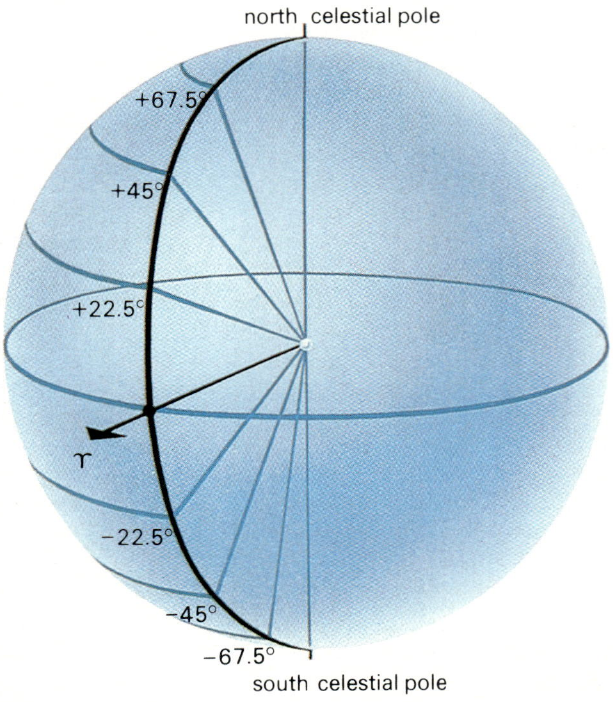

Below left: declination is measured north, or south, of the equator.

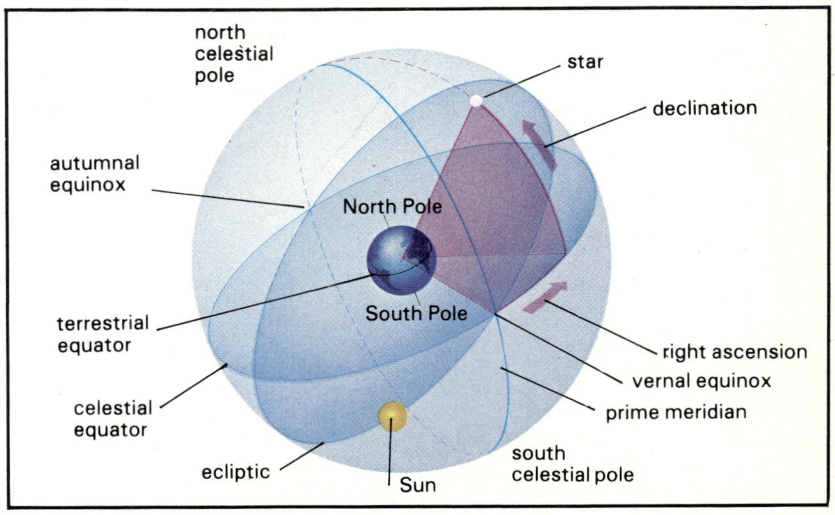

Below: the position of any heavenly body can be given by means of Right Ascension and Declination.

Star charts

The way in which an observer becomes familiar with the sky will naturally depend upon whether he or she is north or south of the equator. It is usually best to start with the circumpolar stars. The star charts on the preceding page and these two pages show the two circumpolar regions, and four equatorial charts arranged by the northern seasons when they are best seen in the evening. Naturally with the rotation of the Earth many other constellations will become visible to those watching throughout the night, and these will be the 'later' stars. It is for this reason that the seasonal maps are arranged 'back-to-front' across these pages.

In the north, the seven bright stars of the Plough (also called the Big Dipper, and many other names), part of the constellation of Ursa Major (the Great Bear), are familiar to many non-astronomers. From this it is easy to find Polaris (the Pole Star) in Ursa Minor, by following the two Pointers Merak and Dubhe, and from there on to the 'W' of Cassiopeia. From these constellations it is easy to begin to work around the pole and down towards the equator.

Lower down, other useful markers are the Milky Way itself, and the constellations of Leo, Scorpius and Pegasus, as well as the 'Summer Triangle' formed by the brightest stars of Cygnus, Lyra and Aquila, and the winter grouping of Orion, Gemini and brilliant Sirius.

In the south the usual starting point is Crux (the Southern Cross) and the two bright stars of Alpha and Beta Centauri. The False Cross in Vela and Carina, brilliant Canopus and the two Magellanic Clouds provide further guides to the other constellations.

The brightest planets frequently distort the patterns of the constellations in which they appear, and may cause some confusion. However, their positions are given in many handbooks and almanacs, as well as many newspapers, and with a little experience they can be reasonably easily recognized.

Below left: **Orion dominates this region of the sky, with M42, the beautiful Orion Nebula. The Pleiades and Hyades are to the north-east, below Algol, an important eclipsing binary. Praesepe, M44, in Cancer is a striking open cluster. Sirius is the brightest star in the sky.**

Below right: **the Square of Pegasus (which includes one star from Andromeda) is easy to locate. M31, the nearest large galaxy, is visible to the naked eye under clear skies. Gamma Andromedae is a beautiful double star. Mira, in Cetus, is sometimes bright enough to be seen without a telescope.**

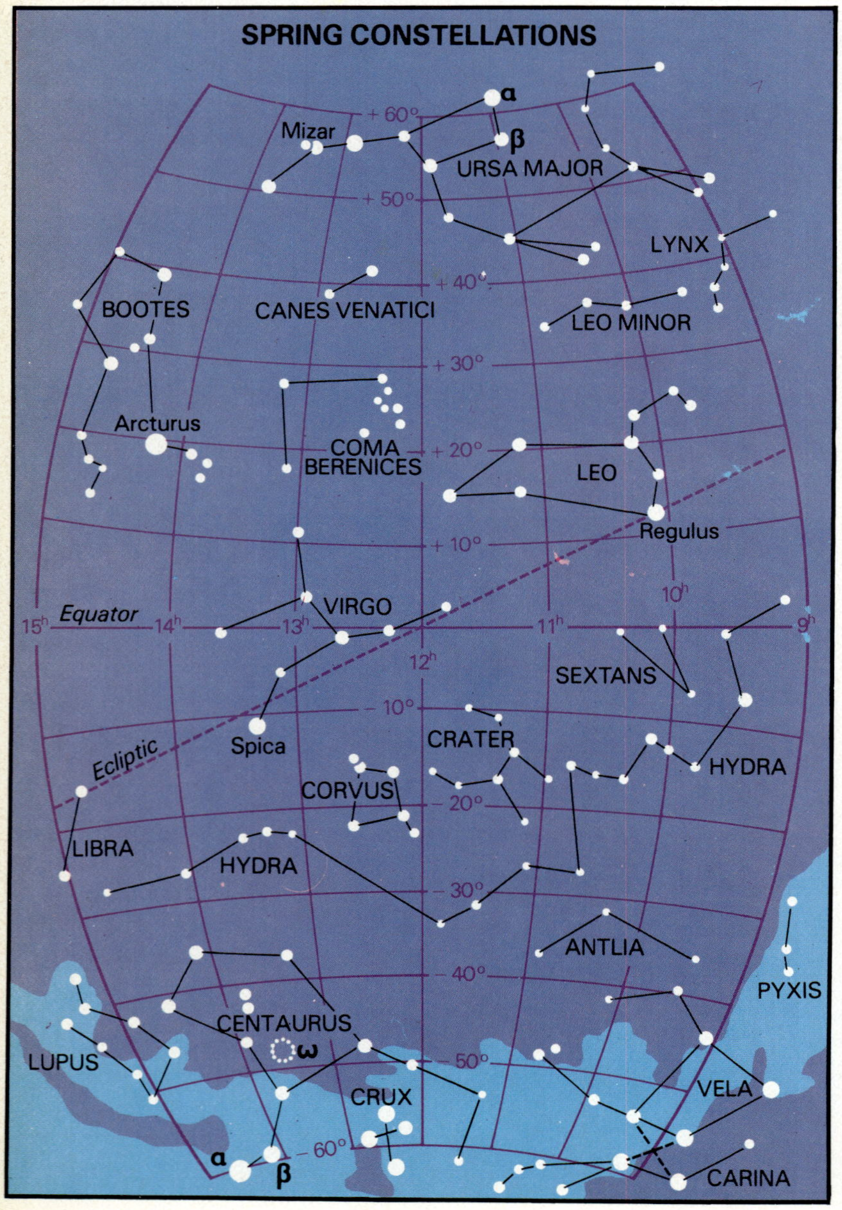

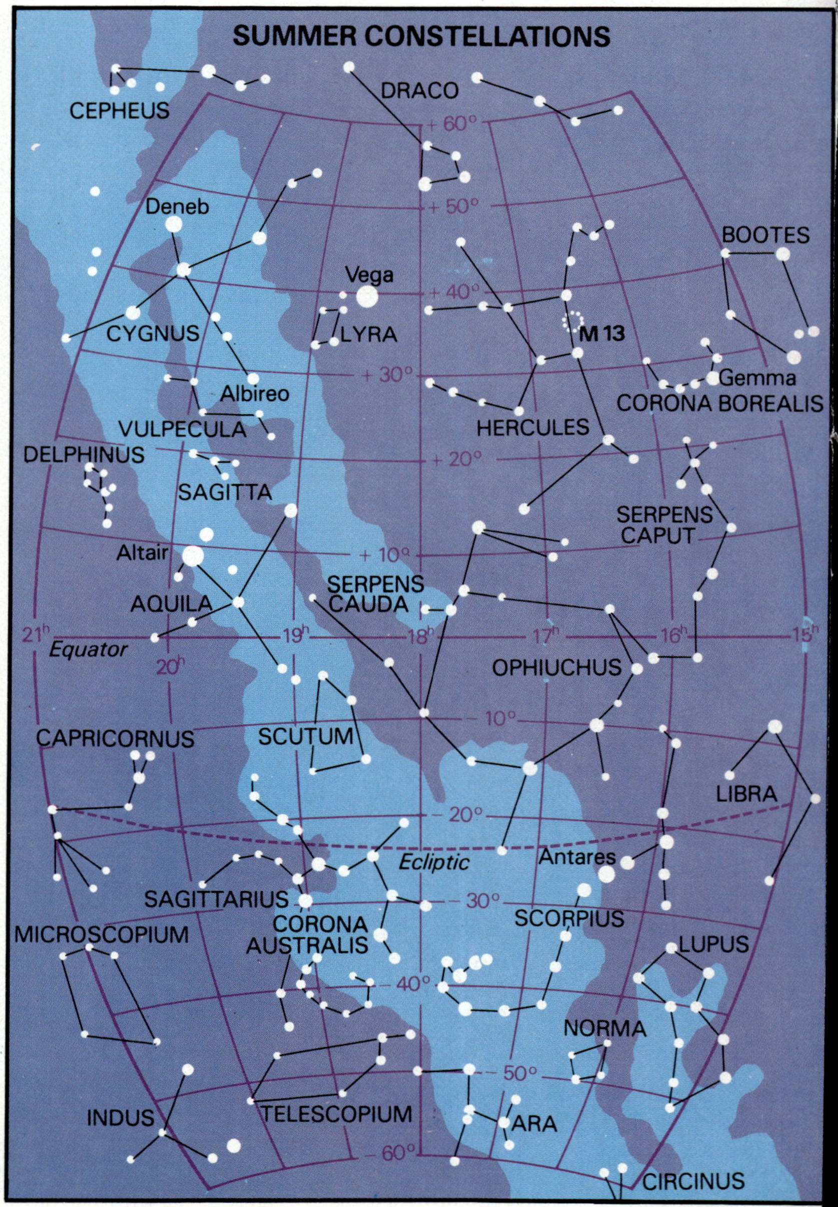

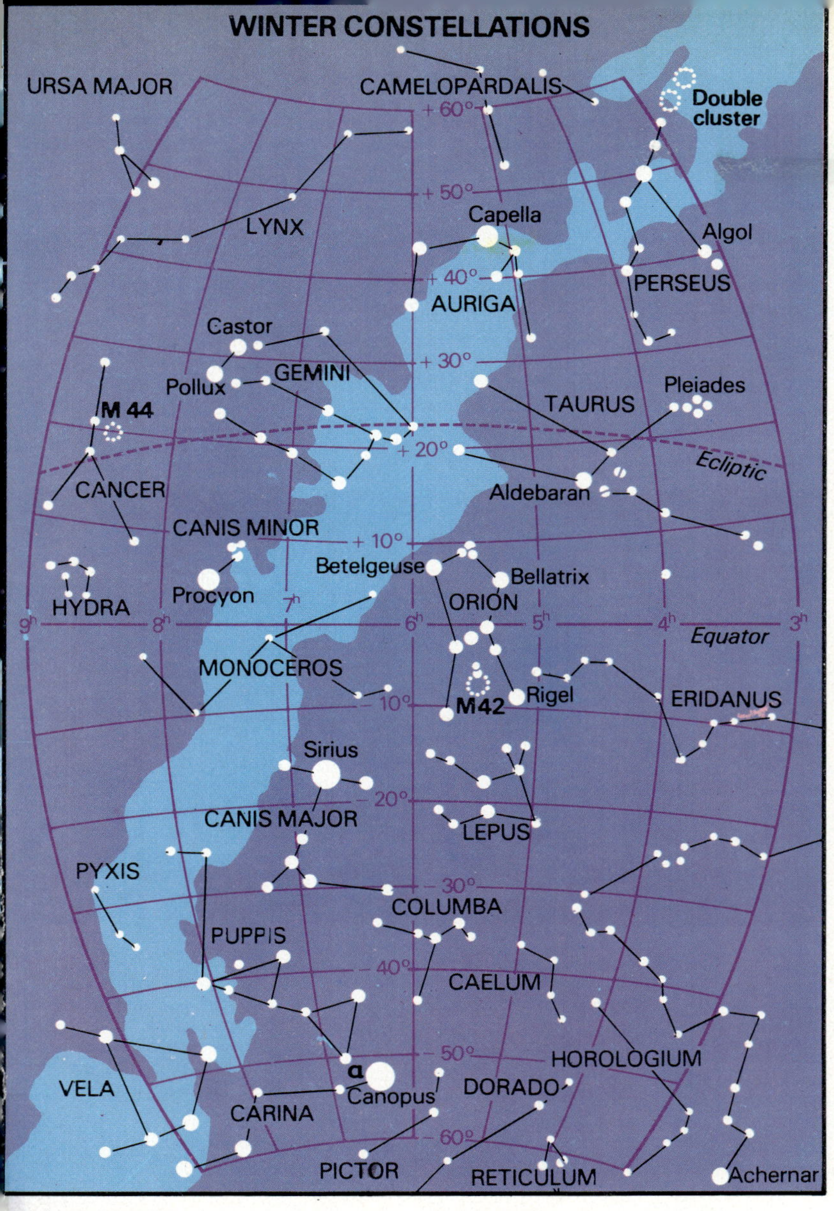

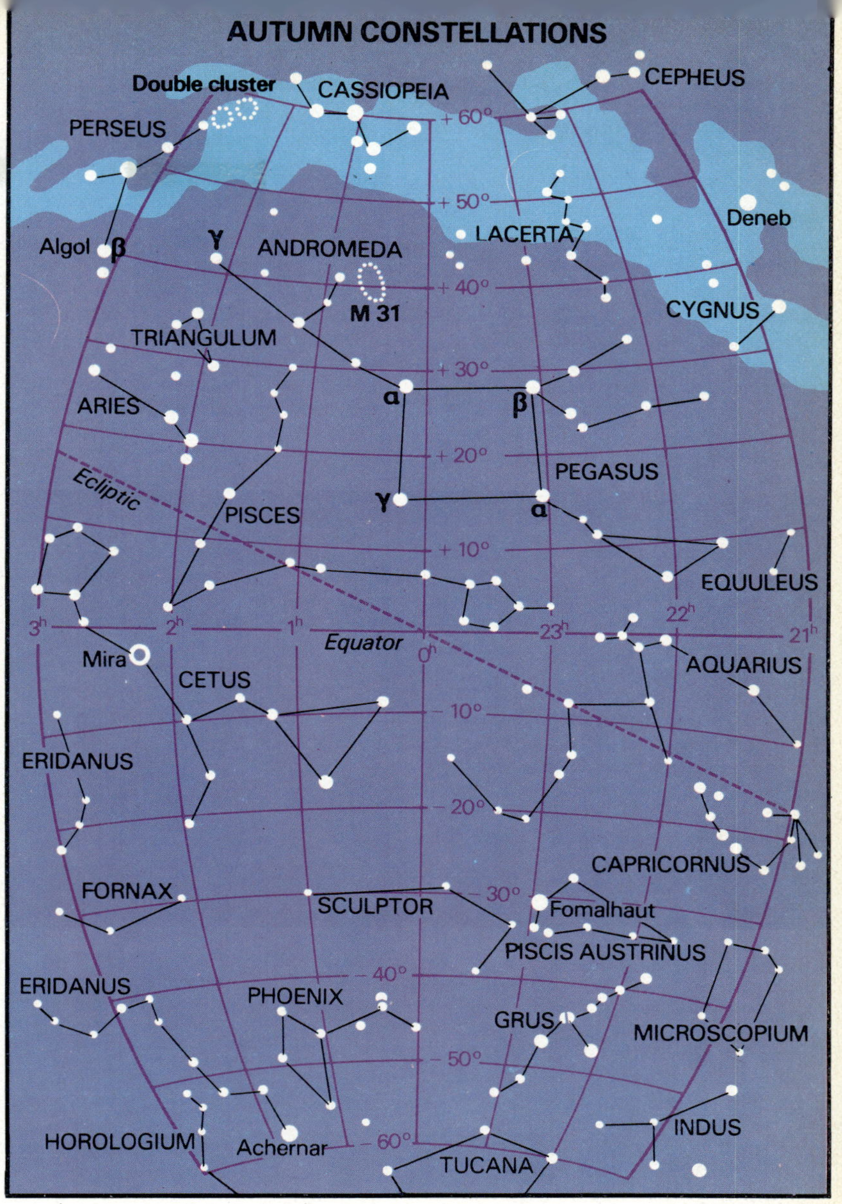

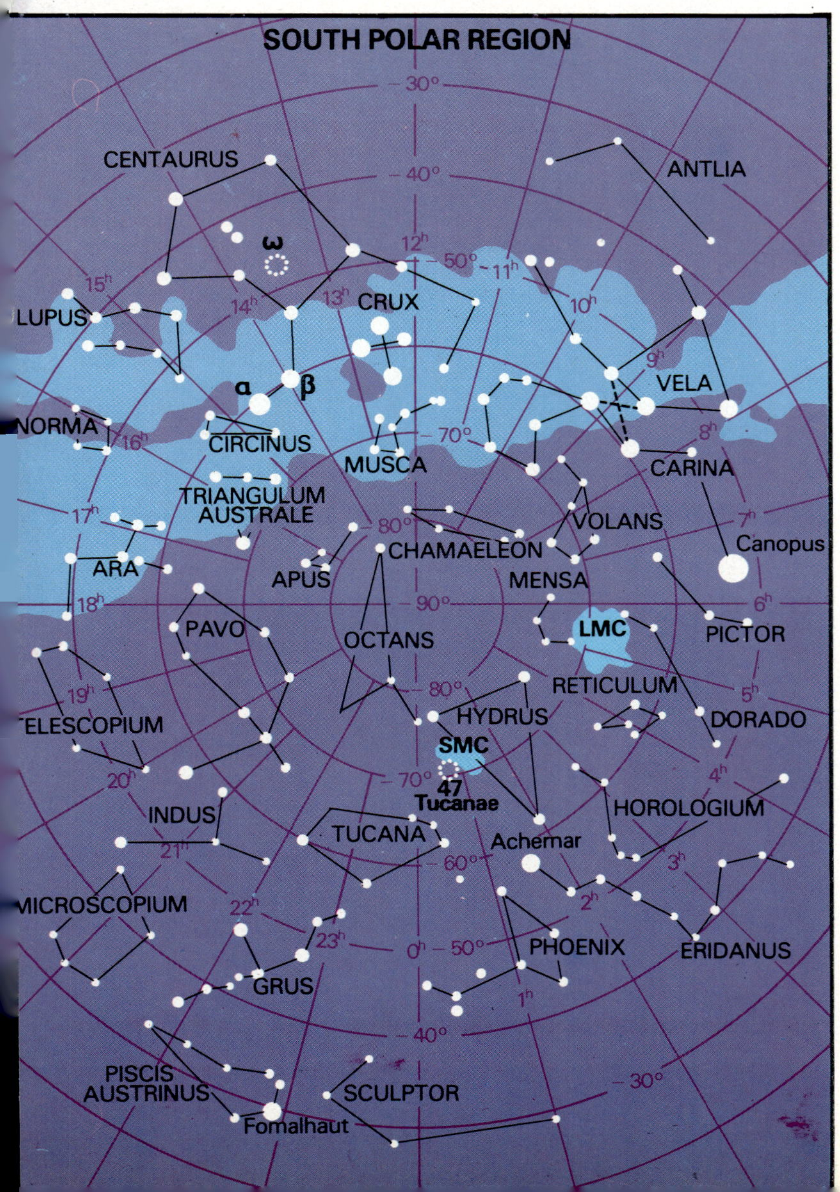

Above left: the 'Summer Triangle' of Deneb, Vega and Altair is easy to recognize. Albireo is another beautiful double star and lies alongside the Great Rift in Cygnus. In Hercules, M13 is the brightest northern globular cluster. Giant Antares lies to the south.

Above: Mizar, in Ursa Major, is a double star visible to the naked eye. Bright Regulus, Spica and Arcturus are the landmarks for this region of the sky. Far to the south Centaurus and Crux may be visible.

Left: the southern polar region. Crux, Alpha and Beta Centauri, Canopus, Achernar and Fomalhaut act as guides in this part of the heavens. The Large (LMC) and Small (SMC) Magellanic Clouds look like detached parts of the Milky Way. The globular clusters 47 Tucanae and Omega Centauri are particularly striking, as is the 'Coalsack' south-east of Crux.

Below: M27, the Dumb-bell Nebula, is a beautiful planetary nebula in Vulpecula.

Glossary

altazimuth A design of telescope mounting which has one motion parallel to the horizon and one at right angles to it.
annular eclipse An eclipse where a ring of the Sun remains visible around the Moon.
astronomical unit The average distance of the Earth from the Sun (149 597 870km).
barred spiral A spiral galaxy where the arms arise from a bar across the centre.
Big Bang The theory that the Universe arose from a single point in time and space.
binary system Two stars which are gravitationally bound to one another.
black dwarf An extinct star at the end of its life. Assumed to exist, but none have yet been discovered.
black hole A region of space surrounding a completely collapsed body – a singularity (q.v.) – from which not even light or other radiation can escape.
Cepheid variable A bright (supergiant) star having very regular changes in magnitude, and which can be used for distance determinations.
cluster A group of stars which have a common origin and appear in the same region of space.
corona The tenuous, outermost layer of the Sun's atmosphere.
cosmic rays Very energetic particles, some accelerated by Jupiter's magnetic field, others produced by novae and supernovae within the Galaxy.
Doppler effect The change in wavelength of light or other radiation due to the relative motion of the source and receiver.
dwarf Actually, an average-sized star for its particular mass (except black or white dwarfs). The Sun is a dwarf star.
ecliptic The Sun's apparent path against the stars.
elliptical galaxy A galaxy which consists of a smooth, more or less flattened system of stars, with little interstellar dust or gas.
equatorial A form of telescope mounting where one axis is parallel to that of the Earth, giving motion in the plane of the celestial equator, and the second axis is at right angles to it.
fireball A brilliant meteor caused by a fairly large body hitting the Earth's atmosphere. May give rise to a meteorite.
gas giant One of the four planets, Jupiter, Saturn, Uranus or Neptune, which mostly consist of various gases.
geocentric Centred on the Earth.
globular cluster A spherical system of stars, formed early in the history of a galaxy.
gravitational waves Variations in the gravitational field produced by motions or collapse of massive bodies.
heliocentric Centred on the Sun.
infrared radiation Heat radiation just beyond the red end of the visible portion of the spectrum.
inner planet One of the four innermost planets in the Solar System, Mercury, Venus, Earth or Mars, often known as the terrestrial planets.
irregular galaxy A galaxy of little apparent form.
kiloparsec 1000 parsecs.
light-year The distance travelled by light in one year (approximately 9·5 million, million km, or 63 240 astronomical units).
lunar eclipse Obscuration of the Moon by the shadow of the Earth.
magnitude The value of the brightness of a star. For historical reasons brighter stars have smaller magnitude values. A star of first magnitude is 100 times as bright as one of sixth magnitude.
mare (pl maria) The fairly flat lava plains upon the Moon.
megaparsec 1 million parsecs.
meteor A tiny, dust-like particle which burns up in the Earth's atmosphere.
meteorite A body which has survived its passage through the Earth's atmosphere.
minor planet One of the many tiny bodies orbiting the Sun, some of which are fragments of larger objects.
nebula (pl nebulae) A cloud of interstellar gas or dust, the former bright when ionised by starlight, the latter bright in a reflection nebula, or dark when obscuring the light of other stars.
neutrino Energetic (massless) particle produced by nuclear reactions, which only rarely reacts with any form of matter.
neutron One of the particles (without electrical charge) which may be regarded as making up the nucleus of atoms.
neutron star The remnant of a supernova explosion, so compressed that it consists almost entirely of neutrons.
nova (pl novae) A star which suddenly undergoes a great increase in brightness – typically ten thousand-fold or more.
parsec Short for parallax second. The distance at which one astronomical unit subtends the angle of one second of arc. A star at this distance would show an apparent movement of 2 seconds of arc against the background stars. (3·2616 light-years or 206 265 astronomical units.)
partial eclipse An eclipse where only part of the Sun (or Moon) is obscured.
planetary nebula A shell of gas shed by an evolving star.
polar co-ordinates A system of co-ordinates for specifying positions in the sky, based upon the celestial equator, and the counterpart of latitude and longitude on Earth.
prominence A cloud of gas seen on the edge of the Sun.
proton One of the particles (with a single positive charge) which may be regarded as making up the nucleus of atoms.
protostar A collapsing mass of dust and gas which will become a true star when nuclear reactions begin to take place in the centre.
pulsar A rapidly-rotating neutron star producing pulsating radio signals.
quasar Highly-energetic objects far distant in space, now thought to be the violent centres of young galaxies.
radio telescope A large aerial (or aerials) which can detect radiation at radio wavelengths from astronomical sources.
red giant A star which has burnt much of its hydrogen and started to evolve rapidly, swelling in size and becoming cooler.
red supergiant An enormous star, hundreds of times larger than the Sun. Antares, a red supergiant, would engulf the orbit of Mars, if placed in the Solar System.
refractor A telescope which uses a main lens to gather the light from an astronomical object.
reflector A telescope using a mirror to concentrate light and produce an image.
retrograde rotation Clockwise rotation as viewed from above the north pole. Opposite to the more usual direct rotation.
singularity An infinitesimal point where our laws in physics must break down. The result of uninterrupted collapse of a massive stellar remnant, and surrounded by a black hole.
solar eclipse Obscuration of the Sun by the body of the Moon.
solar flares Sudden short-lived brightenings of regions of the Sun.
solar wind A flow of charged particles from the Sun.
spectrum The range of wavelengths covered by visible light, and more specifically, the complete range of electromagnetic radiation from gamma-rays (very short) to radio waves (long).
spiral galaxy A galaxy where the disc has distinct spiral arms.
sunspot An apparently darker region of the Sun's surface.
supergiant A very massive star much greater than the Sun in size.
supernova (pl supernovae) A star which undergoes a tremendous explosion – increasing in brightness by ten million-fold or more – and throws out heavy elements into space.
tektite A meteorite-like object found on Earth and probably produced by meteoritic impact.
terrestrial planets The four Earth-like inner planets.
ultraviolet radiation Radiation which is just beyond the blue end of the visible spectrum.
white dwarf The dense remnants of a star nearly at the end of its life.

Bibliography

Books

Amateur Astronomy, C. A. Ronan (ed.), Newnes Books, London, 1984
Astronomy: A Dictionary of Space and the Universe, I. Nicolson, Arrow, London, 1977
Constellations, J. Klepesta and A. Rukl, Hamlyn, London, 1969
Deep Space, C. A. Ronan, Pan, London, 1983
The Encyclopedia of Space Travel and Astronomy, J. Mann (ed.), Octopus, London, 1979
The Guinness Book of Astronomy Facts and Figures, P. Moore, Guinness Superlatives, London, 1979
Make Your Own Telescope, R. Spry, Sidgwick and Jackson, London, 1978
Moon, Mars and Venus, A. Rukl, Hamlyn, London, 1975
The New Solar System, (2nd edn, eds J. K. Beatty, B. O'Leary and A. Chaikin), Cambridge University Press, 1983
Norton's Star Atlas, (17th edn, ed. G. Satterthwaite), Gall Inglis, Edinburgh, 1978
The Young Astronomer's Handbook, I. Ridpath, Hamlyn, London, 1981
Practical Amateur Astronomy, (4th edn, ed. P. Moore), Lutterworth Press, Guildford, 1975
The Practical Astronomer, C. A. Ronan, Pan, London, 1981
Newnes Pocket Guide to Astronomy, M. Chartrand and H. Wimmer, Newnes Books, London, 1984
Yearbook of Astronomy, P. Moore (ed.), Sidgwick and Jackson, London (Annually)

Journals

Astronomy, AstroMedia Corporation, Milwaukee (Monthly)
Journal of the British Astronomical Association, London (Bi-monthly)
Popular Astronomy, Junior Astronomical Society, London (Bi-monthly)
Sky and Telescope, Sky Publishing Corporation, Cambridge, Mass. (Monthly)

Index

Page numbers in *italic* refer to illustrations